PROC REPORT by Example
Techniques for Building Professional Reports Using SAS®

Lisa Fine

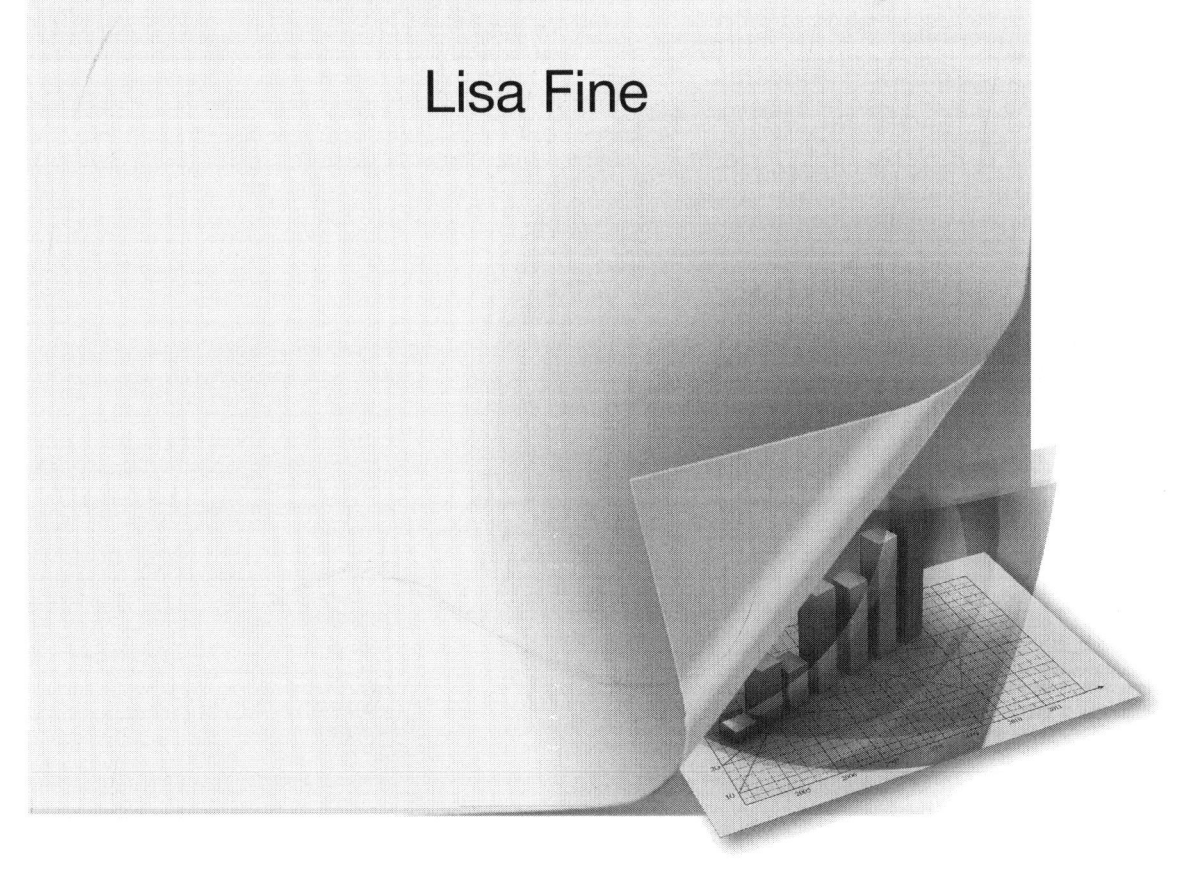

§.sas®

support.sas.com/bookstore

The correct bibliographic citation for this manual is as follows: Fine, Lisa. 2013. *PROC REPORT by Example: Techniques for Building Professional Reports Using SAS®*. Cary, NC: SAS Institute Inc.

PROC REPORT by Example: Techniques for Building Professional Reports Using SAS®

Copyright © 2013, SAS Institute Inc., Cary, NC, USA

ISBN 978-1-61290-784-0

All rights reserved. Produced in the United States of America.

For a hard-copy book: No part of this publication may be reproduced, stored in a retrieval system, or transmitted, in any form or by any means, electronic, mechanical, photocopying, or otherwise, without the prior written permission of the publisher, SAS Institute Inc.

For a web download or e-book: Your use of this publication shall be governed by the terms established by the vendor at the time you acquire this publication.

The scanning, uploading, and distribution of this book via the Internet or any other means without the permission of the publisher is illegal and punishable by law. Please purchase only authorized electronic editions and do not participate in or encourage electronic piracy of copyrighted materials. Your support of others' rights is appreciated.

U.S. Government License Rights; Restricted Rights: The Software and its documentation is commercial computer software developed at private expense and is provided with RESTRICTED RIGHTS to the United States Government. Use, duplication or disclosure of the Software by the United States Government is subject to the license terms of this Agreement pursuant to, as applicable, FAR 12.212, DFAR 227.7202-1(a), DFAR 227.7202-3(a) and DFAR 227.7202-4 and, to the extent required under U.S. federal law, the minimum restricted rights as set out in FAR 52.227-19 (DEC 2007). If FAR 52.227-19 is applicable, this provision serves as notice under clause (c) thereof and no other notice is required to be affixed to the Software or documentation. The Government's rights in Software and documentation shall be only those set forth in this Agreement.

SAS Institute Inc., SAS Campus Drive, Cary, North Carolina 27513-2414.

December 2013

SAS provides a complete selection of books and electronic products to help customers use SAS® software to its fullest potential. For more information about our offerings, visit **support.sas.com/bookstore** or call 1-800-727-3228.

SAS® and all other SAS Institute Inc. product or service names are registered trademarks or trademarks of SAS Institute Inc. in the USA and other countries. ® indicates USA registration.

Other brand and product names are trademarks of their respective companies.

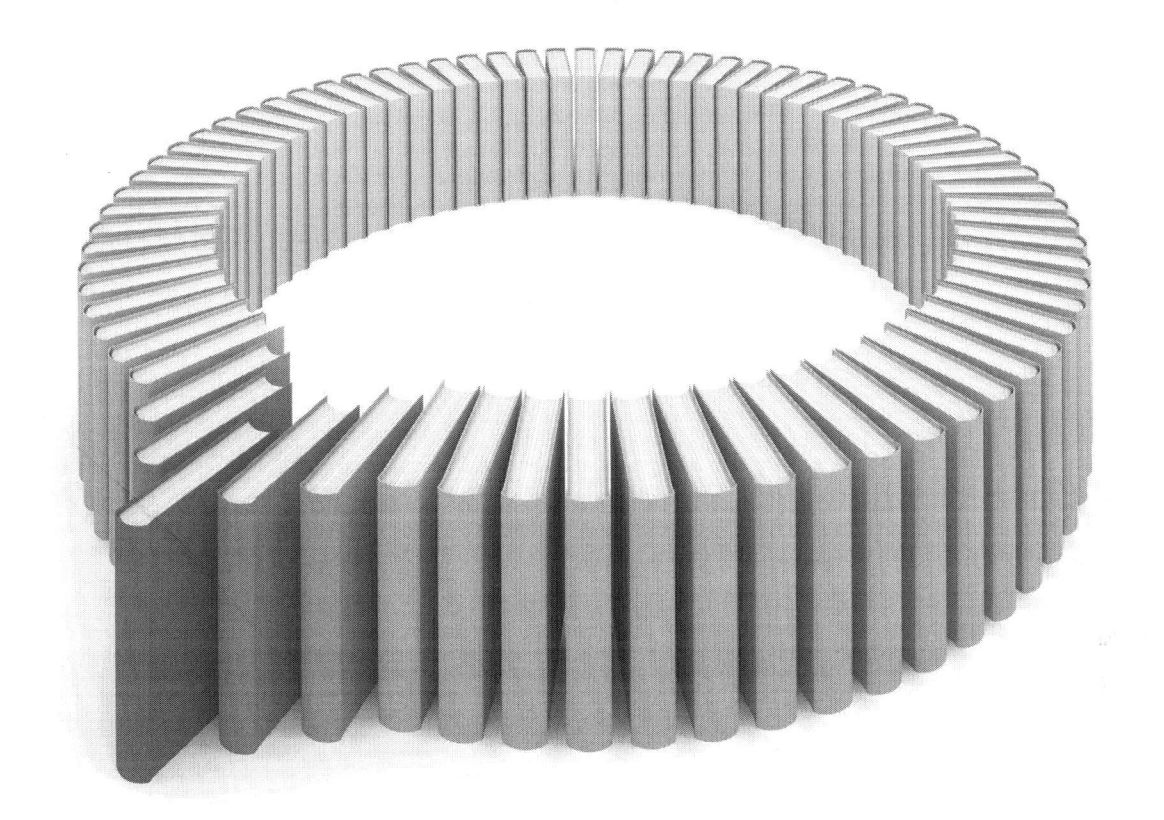

Gain Greater Insight into Your SAS® Software with SAS Books.

Discover all that you need on your journey to knowledge and empowerment.

support.sas.com/bookstore
for additional books and resources.

§.sas.
THE POWER TO KNOW.

SAS and all other SAS Institute Inc. product or service names are registered trademarks or trademarks of SAS Institute Inc. in the USA and other countries. ® indicates USA registration. Other brand and product names are trademarks of their respective companies. © 2013 SAS Institute Inc. All rights reserved. S107969US.0413

Contents

About This Book xi
Acknowledgements xvii
Chapter 1: Creating Complementary Reports 1
Introduction 2
Example: Department Store Summary and Detail Reports 2
Goals for Creating Complementary Reports 4
    Key Steps 5
Source Data 6
ODS Style Template Used 7
Programs Used 8
Implementation 8
    Create a Setup Program that Contains Common SAS Code 8
    Ch1Setup.SAS 8
Writing the Detail Report Program 14
    Detail Report Pre-Processing Code 15
Detail Report: Titles, Footnotes, and ODS RTF Preparation 15
    Code for Titles, Footnotes, and ODS RTF Preparation 16
Producing the Report with PROC REPORT 17
    Detail Report - PROC REPORT Code 19
Writing the Summary Report Program 23
    Map Separate Variables/Values to One Column for PROC REPORT 23
    Summary Report - Pre-Processing Code 24
    Assign Report Order to Variables 27
    Code for Creating Ordered Variables 28
    Summary Report: Titles, Footnotes, and ODS RTF Preparation 30
    Code for Titles, Footnotes, and ODS RTF Preparation 30
Producing the Report with PROC REPORT 31
    Summary Report - PROC REPORT Code 34

Chapter 1 Summary 40
Chapter 2: Formatting Highly Detailed Reports 41
Introduction 42
Example: Format National Sales Report.. 42
Goals for Formatting the National Sales Report... 45
    Key Steps... 45
Source Data 46
ODS Style Template Used 46
Programs Used... 47
PROC TEMPLATE Program to Create New Style Template 47
The "Before Formatting" Program (Program 2.1) 49
Implementation 53
Transforming Figure 2.1 Into Figure 2.2.. 53
Displaying Region as a Line Above Each Report Page.. 53
    Overview of Region Display... 53
    Code to Make the Region Display in Figure 2.3 54
Displaying Store and Branch Column Data in Bold Blue Font 54
    Code for Store and Branch Display 55
How to Insert Arrows for Quick Reference to Sales Increases/Decreases........... 56
    Overview on Arrow Insertion 56
    Code for Arrow Insertion 56
How to Add Spanning Headers, Bottom Cell Borders, and Underlines 59
    Highlights on Adding Spanning Headers, Borders, and Underlines 59
    Code for Adding Spanning Headers, Borders, and Underlines 60
Adding Blank Columns to Make the Report More Legible 62
    Overview of Adding Blank Columns 62
    Code for Adding Blank Columns... 62
Style: Add a Blank Line After Each Summary Line 65
    Highlights on Styling Summary Line and Adding a Blank Line 65
    Code for Styling Summary Line and Adding a Blank Line............................ 65
Chapter 2 Summary 66
Chapter 3: Reporting Different Metrics Within a Column............................. 69
Introduction 70
Example: Demographic and Baseline Characteristics Report............................... 70
Goals for the Demographics and Baseline Characteristics Report....................... 72

Key Steps	72
Source Data	73
ODS Style Template Used	74
Programs Used	74
Implementation	74
Obtain Population Counts for Column Headers and Denominators	74
Code for Obtaining Population Counts	75
Categorical Variables: Obtain Counts and Percentages	75
Code for Obtaining Categorical Counts and Percentages	76
Continuous Variables: Descriptive Data	81
Macro Code for Obtaining Descriptive Statistics	81
Create Final Table: Combine TABULATE and MEANS Results	85
Code for Combing the Results	85
Produce the Report via PROC REPORT	89
PROC REPORT Code	89
Chapter 3 Summary	91
Chapter 4: Lesion Data Quality Report—COMPUTE Blocks	**95**
Introduction	96
Example: Lesion Data Quality Report	96
Goals for Creating the Lesion Data Quality Report	98
Key Steps	99
Source Data	99
ODS Style Template Used	100
Programs Used	101
Implementation	101
COMPUTE Block Variables: DATA Step (Temporary) Versus REPORT (COLUMN Statement) Variables	101
ORDER by and Print Subject ID on Every Row with Greying Font	102
Program for Subject ID Display	102
Identify Potential Data Issues	107
Code for Displaying Potential Data Issues	108
Final Formatting: Create Spanning Headers	119
Chapter 4 Summary	120
Chapter 5: Multi-Sheet Workbook With Histograms—ExcelXP Tagsets Report	**123**

Introduction 124
Example: Multi-Sheet Workbook Containing Heart Study Results........................ 124
Goals for Creating the Multi-Sheet Workbook 128
    Key Steps.. 128
Source Data 129
ODS Style Template Used 130
Programs Used 134
Implementation 134
Create Formats and Informats.. 134
    Code for Creating Formats and Informats... 134
Obtain Counts and Percentages 137
    Code for Obtaining Counts and Percentages... 137
Producing the Workbook With PROC REPORT and ODS Tagset........................ 141
    Code for Opening, Closing, and Setting Initial Options for the ExcelXP Workbook 142
Producing the Specific Worksheets 144
    Code for Producing ByStatusCOL and ByStatusROW Worksheets................ 144
    Code for Producing ByStatusALL Worksheet 149
Chapter 5 Summary 154

Chapter 6: Using the ACROSS Option to Create a Weekly Sales Report.. 155
Introduction 156
Example: Weekly Sales Report 156
Goals for Creating a Weekly Sales Report 158
    Key Steps.. 158
Source Data 158
ODS Style Template Used 160
Programs Used 160
Implementation: Creating the ODS Style Template 160
    Proc Template Code.. 160
Obtain Calendar Grid and Merge With Sales 162
Produce the Report 166
    Code for Producing the Report... 167
Place Holders for Data Not Yet Available 177
Chapter 6 Summary 179

Chapter 7: Embedding Images in a Report... 181
Introduction 182

Example: Tables Displaying Iris Flower Measurements	182
Goals for Embedding Images in Reports	188
Source Data	188
ODS Style Templates Used	190
Programs Used	190
Implementation	190
Setup Options, File Paths, and Image File Names	190
Program Setup Code	191
Example 1: Obtain Images as Column of Data	192
Code for Obtaining Images as Column of Data	193
Example 2: Repeated Images Above and Below Table	197
Code for Repeating Images Above and Below Table	198
Produce the Report	200
Example 3: Display Images as Column Headers	203
Code for Displaying Images as Column Headers	203
Example 4: Display Image in Page Title	206
Code for Displaying Images in Page Titles	207
Example 5: Display Image Above Body of Table	208
Code for Displaying Image Above Body of Table	210
Example 6: Display Watermark on Report	212
Chapter 7 Summary	213
Chapter 8: Combining Graphs and Tabular Data	**215**
Introduction	216
Example: Dashboard Report of Shoe Sales	216
Goals for Creating the Shoe Sales Dashboard	218
Key Steps	218
Source Data	218
ODS Style Template Used	219
Programs Used	219
Implementation	220
Create a Summary Data Set using PROC REPORT	220
Code for Creating a Summary Data Set	220
Obtain Regional Ranking Information	222
Code for Obtaining Regional Ranking Information	222
Create a New ODS Style Template	223

Create the ODS LAYOUT for the Report 226
Create Formats Needed for Outputs 226
Use PROC SGPLOT to Create Vertical Bar Charts 227
    Code for SGPLOT Vertical Bar Charts 227
Using PROC SGPLOT to Create a Horizontal Bar Chart 230
    Horizontal Bar Chart Code 230
Using PROC REPORT to Obtain Tabular Output 231
Using PROC SGPANEL to Create Bar Charts for the Top 3 Regions 232
Chapter 8 Summary 235

Chapter 9: Using PROC REPORT to Obtain Summary Statistics for Comparison 237

Introduction 238
Example: Vehicle MSRP Comparison Report 238
Goals for MSRP Comparison Report 240
    Key Steps 240
Source Data 240
ODS Style Template Used 242
Programs Used 242
Implementation 242
Initial PROC REPORT for Obtaining Statistics 242
    Code for Obtaining Statistics 242
Produce the Report 245
    Code for Print Report 245
Chapter 9 Summary 254
References 255
DATA SETS 261
Index 263

About This Book

Purpose
Personally, I learn best through experience and working through real programming examples. I wrote this book to provide PROC REPORT users with step-by-step instructions for turning data sets into final reports. The examples provide solutions to common programming challenges that are encountered during the production of various types of reports.

Is This Book for You?
This book is for PROC REPORT users who want to learn more about PROC REPORT and ODS features by working through examples. Each chapter demonstrates how to produce a different type of report. The book is applicable to SAS users from all disciplines.

Prerequisites
SAS programmers that are familiar with BASE SAS, basic SAS procedure syntax and logic, and have some familiarity with PROC REPORT are likely to benefit most from this book.

About the Examples

Software Used to Develop the Book's Content
This book was written using the following operating system and products:

Operating System: Microsoft® Windows® Workstation for x64, Windows 7 Home Premium

Products: BASE SAS and SAS/STAT: SAS 9.3 TS1M2, Rev. 930_12w37

SAS/GRAPH: SAS 9.3 TS1M2, Rev. 930_12w50

Example Code and Data
Visit the author's page at http://support.sas.com/publishing/authors/fine.html to access the data and programs used in this book.

The following table describes the Program Name(s) and Data Set Name(s) that correspond to each chapter.

Chapter	Program Name(s)	Data Set Name(s)
Chapter 1	Ch1Setup	CH1STORE
	Ch1Detail	CH1ECOMM
	Ch1Summary	
Chapter 2	Ch2Format	CH2SALES
Chapter 3	Ch3Demo	CH3DEMO
Chapter 4	Ch4qc	CH4LESN
		Chapters 5 through 9 use SAS-provided data sets. The following data sets have been copied from the SAS 9.3 SASHELP library to the author's webpage with permission of the SAS Institute: Copyright 2013, SAS Institute Inc., Cary, NC, USA. All Rights Reserved. Reproduced with permission of SAS Institute Inc., Cary, NC.
Chapter 5	Ch5Tgxml	SASHELP.HEART
Chapter 6	Ch6Cal	SASHELP.SNACKS
Chapter 7	Ch7Images	SASHELP.IRIS
Chapter 8	Ch8Graph	SASHELP.SHOES
Chapter 9	Ch9Stat	SASHELP.CARS

For SAS 9.1.3 Users

This book was written based on SAS 9.3. The author's page at http://support.sas.com/publishing/authors/fine.html provides some SAS 9.1.3 code workarounds for the chapter examples that use SAS 9.2 features. For the SAS 9.1.3 user who wants to learn more about what's new in SAS 9.2, the following list summarizes key SAS 9.2 enhancements that are relevant to this book:

SAS 9.2 Key Enhancements Relevant to Book

- PROC TEMPLATE is simplified

- A new PROC REPORT feature, SPANROWS, is introduced

- A new output delivery system, ODS.TAGSETS.RTF is introduced (provides Continued note and Watermark options)

- New BORDER Control elements and additional style control with the TEXTDECORATION style element

- PDFTOC (ability to collapse a PDF table of contents)

- The Unicode inline style function allows direct insertion of Unicode characters

- Allowable character length is increased

The author suggests the following references for a summary of what is new in SAS 9.2.

Booth, Allison McMahill. 2011. "Beyond the Basics: Advanced REPORT Procedure Tips and Tricks Updated for SAS® 9.2." Proceedings of the 2011 SAS Global Forum - Paper 246-2011, Cary, NC: SAS Institute Inc. Available at http://support.sas.com/resources/papers/proceedings11/246-2011.pdf

Huntley, Scott. 2006. "Let the ODS PRINTER Statement Take Your Output into the Twenty-First Century." Proceedings of the 31st Annual SAS Users Group International Conference - Paper 227-31. Cary, NC: SAS Institute Inc. Available at http://www2.sas.com/proceedings/sugi31/227-31.pdf.

SAS Institute Inc. 2009. KNOWLEDGE BASE / SAMPLES & SAS NOTES. "Usage Note 15883: Length limitations when submitting SAS code." Cary, NC: SAS Institute Inc., Available at http://support.sas.com/kb/15/883.html.

SAS Institute Inc. 2013. KNOWLEDGE BASE FOCUS AREAS. "Base SAS Enhancements to ODS RTF for SAS 9.2." Cary, NC: SAS Institute Inc., Available at http://support.sas.com/rnd/base/new92/92rtf.html.

SAS Institute Inc. 2013. KNOWLEDGE BASE / SAMPLES & SAS NOTES. "Sample 49590: Insert special symbols as a table value in ODS MARKUP destinations." Cary, NC: SAS Institute Inc., Available at http://support.sas.com/kb/49/590.html.

Smith, Kevin D. 2006. "The TEMPLATE Procedure Styles: Evolution and Revolution." Proceedings of the 31st Annual SAS Users Group International Conference - Paper 053-31. Cary, NC: SAS Institute Inc. Available at http://www2.sas.com/proceedings/sugi31/053-31.pdf.

For an alphabetical listing of all books for which example code and data is available, see http://support.sas.com/bookcode. Select a title to display the book's example code.

If you are unable to access the code through the website, send an e-mail to saspress@sas.com.

Output and Graphics

Each chapter discusses how the output and graphics were created for each specific example.

While much of the output and graphics were created in color, the printed SAS book version displays images in black and white. Color images are available from the author's web page at http://support.sas.com/publishing/authors/fine.html. **The e-book version is available in color.**

Additional Resources

SAS offers you a rich variety of resources to help build your SAS skills to explore and apply the full power of SAS software. Whether you are in a professional or academic setting, we have learning products that can help you maximize your investment in SAS.

Bookstore	http://support.sas.com/bookstore/
Training	http://support.sas.com/training/
Certification	http://support.sas.com/certify/
SAS Global Academic Program	http://support.sas.com/learn/ap/
SAS OnDemand	http://support.sas.com/learn/ondemand/

Or

Knowledge Base	http://support.sas.com/resources/
Support	http://support.sas.com/techsup/
Training and Bookstore	http://support.sas.com/learn/
Community	http://support.sas.com/community/

Keep in Touch

We look forward to hearing from you. We invite questions, comments, and concerns. If you want to contact us about a specific book, please include the book title in your correspondence.

To Contact the Author Through SAS Press

By e-mail: saspress@sas.com

Via the Web: http://support.sas.com/author_feedback

SAS Books

For a complete list of books available through SAS, visit http://support.sas.com/bookstore.

Phone: 1-800-727-3228

Fax: 1-919-677-8166

E-mail: sasbook@sas.com

SAS Book Report

Receive up-to-date information about all new SAS publications via e-mail by subscribing to the SAS Book Report monthly eNewsletter. Visit http://support.sas.com/sbr.

This book is dedicated to Rick, who is a model of courage and the best big brother I could ever imagine.

Acknowledgements

I would like to express my gratitude to the following individuals:

To Carol Matthews, Ginger Lewis, Paul Slagle, Andy Newcomer, Jeff Parno, James (Jim) Young, and Scott McBride for teaching me the basics of SAS programming for the pharmaceutical industry.

To Lauren Lake, for her kind mentorship at SAS Global Forum (SAS® Global Forum Presenter Mentoring Program). Lauren provided invaluable feedback on my presentations, papers, and on my initial PROC REPORT book proposal.

To Michael Raithel, whose enthusiasm is contagious. Michael convinced me it was not a matter of if, but when, I wrote a book. Because he was so encouraging, I wrote a book!

To Cynthia Zender, for her insight and encouragement to write a "By-Example" book. Cynthia's feedback became the basis for the content of this book.

To Art Carpenter, for his feedback on my initial PROC REPORT book proposal, and whose writings and courses continue to make me a better SAS programmer.

To Russell Lavery, for his generosity in sharing his expertise. Russell suggested readings, talked through examples, and provided constructive feedback that motivated me past roadblocks I encountered during the writing process.

To Greg McCullough of Iris City Gardens, for providing the iris images for the book (Chapter 7).

To my book reviewers whose depth of experience and concise feedback clearly improved the book:

Allison Booth, Robin Crumpton, Mei Du, Jane Eslinger, Tim Hunter, Bari Lawhorn, Russell Lavery, Matt Nizol

To my editors who saw the book through to completion by coordinating the many aspects of book production:

Stephenie Joyner, Brenna Leath, George McDaniel, Shelley Sessoms

To all at the SAS Institute who have provided assistance along the way:

Jennifer Dilley, Candy Farrell, Thais Fox, Shelly Goodin, Robert Harris, Julie Platt, Cindy Puryear

I would also like to express my gratitude to the very special people in my life:

To my parents, for their never ending love and belief in me.

To my husband Mark, and Kailee, Josh and Jessica, for making every day better with their smiles.

To Robin, Rick, Shari, and Marc, for always finding a way to make me laugh, especially at myself.

To Eileen, for remaining a life-long mentor and always teaching by example.

Chapter 1: Creating Complementary Reports

Introduction 2

Example: Department Store Summary and Detail Reports 2

Goals for Creating Complementary Reports 4

    Key Steps 5

Source Data 6

ODS Style Template Used 7

Programs Used 8

Implementation 8

    Create a Setup Program that Contains Common SAS Code 8

    Ch1Setup.SAS 8

Writing the Detail Report Program 14

    Detail Report Pre-Processing Code 15

Detail Report: Titles, Footnotes, and ODS RTF Preparation 15

    Code for Titles, Footnotes, and ODS RTF Preparation 16

Producing the Report with PROC REPORT 17

    Detail Report - PROC REPORT Code 19

Writing the Summary Report Program 23

Map Separate Variables/Values to One Column for PROC REPORT 23

Summary Report - Pre-Processing Code 24

Assign Report Order to Variables 27

Code for Creating Ordered Variables 28

Summary Report: Titles, Footnotes, and ODS RTF Preparation 30

Code for Titles, Footnotes, and ODS RTF Preparation 30

Producing the Report with PROC REPORT 31

Summary Report - PROC REPORT Code 34

Chapter 1 Summary 40

Introduction

As programmers, we're often asked to provide a series of reports that will be used in conjunction with one another instead of simply providing one report to be used in isolation. An example of this is a request to provide a summary report accompanied by a detailed report so that the source of the summary information is known and can be verified.

Example: Department Store Summary and Detail Reports

For example, a department store chain's corporate office may request a Summary report showing store sales by category, along with a corresponding Detail report showing the source of the sales. The Summary report allows the recipient to see the "big picture" of how the corporation is performing. The Detail report allows the reader to better understand the makeup of the summary results and provides the ability to verify that the summary report correctly accounts for the detailed information.

For the purpose of presenting the simpler example first, production of the Detail Report will be presented prior to developing the Summary Report. Figures 1.1 and 1.2 show sample Detail and Summary reports, respectively.

Figure 1.1 Detail

2011 Sales - Detail

Store	*Region*	*Store Type*	*Sales*
1 / Smith-Fine Goods	East	E-Commerce In-Store	$107,024 $200,310 **$307,334**
	West	E-Commerce In-Store	$90,033 $189,356 **$279,389**
	North	E-Commerce In-Store	$88,024 $99,346 **$187,370**
	South	E-Commerce In-Store	$110,009 $107,220 **$217,229**
2 / J.B. Prog & Co.	East	E-Commerce In-Store	$120,050 $89,499 **$209,549**
	West	E-Commerce In-Store	$110,631 $44,250 **$154,881**
	North	E-Commerce In-Store	$97,897 $100,171 **$198,068**
	South	E-Commerce In-Store	$102,425 $79,465 **$181,890**
3 / XYZ Warehouse	East	E-Commerce In-Store	$159,902 $49,126 **$209,028**
	West	E-Commerce	$136,647 **$136,647**
4 / E-Home Store	East	E-Commerce	$200,120 **$200,120**
	West	E-Commerce	$185,600 **$185,600**

Note: Sales are rounded up to the nearest dollar.
Reference: Figure 1.2: Summary Report

Figure 1.2 Summary

2011 Sales - Summary

Category	Sales
Total	$2,467,105
Store	
1 / Smith-Fine Goods	$991,322
2 / J.B. Prog & Co.	$744,388
3 / XYZ Warehouse	$345,675
4 / E-Home Store	$385,720
Region	
East	$926,031
West	$756,517
North	$385,438
South	$399,119
Store Type	
E-Commerce	$1,508,362
In-Store	$958,743

Note: Sales are rounded up to the nearest dollar.
Reference: Figure 1.1: Detail Report

Goals for Creating Complementary Reports

Before getting into the actual implementation of creating complementary reports, it is useful to think about the overall goals and steps needed for achieving these goals.

There are several steps we can implement that will allow for accuracy and ease of use to the reader when analyzing multiple reports. These include:

- A consistent report template
- Consistency of data definitions across reports
- Consistent labels so the end user can easily match items across the reports
- Footnote references clarifying which reports correspond to each other
- Quality checks to ensure that numbers/statistics correspond across reports

Key Steps

For both the Detail and Summary report, three main steps are addressed in the programming: 1) pre-processing the data prior to PROC REPORT, 2) preparing the titles, footnotes, and ODS destination to be used, and 3) using PROC REPORT to create the Detail and Summary reports.

1. **Pre-processing (prior to PROC REPORT)**

    o **Determine the aspects that are common to both Detail and Summary reports**

For this example, the incoming data sets STORE and ECOMM (see Tables 1.1 and 1.2) need to be concatenated, as both Detail and Summary reports display both In-Store and E-commerce sales. The Store Type variable, designating In-Store versus E-Commerce Sales needs to be created since it is not a variable in the existing data sets. The Store Name variable needs to display Store Number before Store Name in the format Store Number / Store Name. Both reports share the title "2011 Sales" and the footnote "Note: Sales are rounded up to the nearest dollar." The Summary and Detail reports should be based on a common style template for a common look. Both reports, while containing different file names, need to be output to the same file path.

    o **Identify the unique pre-processing steps needed for each individual report**

In the case of the Detail report, very little additional processing is needed after the common programming steps are implemented. The data will already be in the structure needed for PROC REPORT. Only one more change is needed, and that is to create a numeric region variable for ordering Region rows in the desired report order.

In the case of the Summary report, Figure 1.2 shows the originally separate variables and their corresponding values Store (STORENAM), Region (REG), and Store Type (TYPE) in one column. The programmer can restructure the data set to map the original variables and values to one variable or column. In addition, a Total row is created as the first row of the report. Because a particular order of rows is desired, variables are created for the specific purpose of assigning numeric order to header and actual data rows. These order variables will be specified later in PROC REPORT as variables that should be used to group and order rows.

2. **Titles, Footnotes, and ODS RTF Preparation**

The title and footnote text common to both Detail and Summary reports will be assigned to macro variables as one of the pre-processing steps. These macro variables will just need to be referenced in the title and footnote section prior to each PROC REPORT section. Additional title and footnote text specific to each report will be added.

You will need to specify the ODS destination before producing the reports. This includes opening the RTF destination and the specific file path and file name to which the report will be created. In addition, the programmer can assign the ODS style (if different than the default template) that should be applied to the reports. A macro variable (named ODSOPT) specifies ODS options such as orientation (Portrait or Landscape), printing, or suppression of SAS SYSTEM information (dates, page numbers) and other options we wish to apply to the reports. This macro variable is later referenced in an OPTIONS statement.

3. **Producing the Report with PROC REPORT**

Once the data is in an appropriate format and the RTF destination is designated, we are ready to produce the reports. The Detail report makes use of PROC REPORT's ORDER option to arrange rows and prevent repeating values, and a BREAK statement with a SUMMARIZE option to sum sales per Store-Region. STYLE options that apply uniquely to the Detail report are specified in the PROC REPORT code, rather than in the PROC TEMPLATE program.

The Summary Report takes advantage of PROC REPORT's GROUP option to consolidate and establish the order of rows, as well as sum the sales per group. The Summary report also makes use of STYLE options specified directly in the PROC REPORT code.

Source Data

There are two source data sets, each having four variables. The data set STORE contains In-Store sales information, and the data set ECOMM contains E-Commerce sales information. The four common variables are:

- STORENUM (Store Number)
- REG (Region responsible for sales)
- STORENAM (Store Name)
- SALES (2011 Sales)

Tables 1.1 and 1.2 display the STORE and ECOMM data, respectively.

Table 1.1 STORE Data

STORENUM	REG	STORENAM	SALES
1	East	Smith-Fine Goods	$200,310
1	West	Smith-Fine Goods	$189,356
2	East	J.B. Prog & Co.	$89,499
2	West	J.B. Prog & Co.	$44,250
1	North	Smith-Fine Goods	$99,346
1	South	Smith-Fine Goods	$107,220
2	North	J.B. Prog & Co.	$100,171
2	South	J.B. Prog & Co.	$79,465
3	East	XYZ Warehouse	$49,126

Table 1.2 ECOMM Data

STORENUM	REG	STORENAM	SALES
1	East	Smith-Fine Goods	$107,024
1	West	Smith-Fine Goods	$90,033
2	East	J.B. Prog & Co.	$120,050
2	West	J.B. Prog & Co.	$110,631
1	North	Smith-Fine Goods	$88,024
1	South	Smith-Fine Goods	$110,009
2	North	J.B. Prog & Co.	$97,897
2	South	J.B. Prog & Co.	$102,425
3	East	XYZ Warehouse	$159,902
3	West	XYZ Warehouse	$136,647
4	East	E-Home Store	$200,120
4	West	E-Home Store	$185,600

ODS Style Template Used

Both reports are produced in the Output Delivery System (ODS) Rich Text Format (RTF) destination starting with the ODS Journal style template, and modifying the template to a new style named JournalR. The ODS style template is specified prior to the PROC REPORT section in the statement

```
ods rtf style=journalR file="{PATH\FILENAME}.rtf";
```

Programs Used

Though this example presents only two small reports, in reality a list of summary reports along with a list of corresponding detail reports may be requested. Therefore, Detail and Summary reports are created in separate SAS programs. In total, three programs are created and used:

- Ch1Setup.SAS
- Ch1Detail.SAS
- Ch1Summary.SAS

Implementation

Create a Setup Program that Contains Common SAS Code

In this step we create a setup program which contains the code to be used in both Detail and Summary reports. The program includes creating a common style template, macro variable assignments, and data manipulation identified as aspects common to both reports.

Specifically, Ch1Setup.SAS will accomplish the following:

- Set up a common ODS Style Template.
- Create Macro Variables to be used in Report Creation.
- Declare the ODS ESCAPE Character.
- Develop a Macro to Modify Incoming Data for Reports.

Ch1Setup.SAS

```
** Program Name: CH1Setup.SAS;
** Modify Journal Style Template; ❶
proc template;
   define style styles.journalR;
```

```
      parent=styles.journal;

   class fonts /

      "DocFont"     = ("Georgia", 9 pt) /*Apply to data in cells*/

      "EmphasisFont" = ("Georgia", 9 pt, Bold) /*Apply to
lines/summaries*/

      "HeadingFont"  = ("Georgia", 16 pt, Bold Italic); /*Apply to
headers*/

   class table /

      rules=none /*Override Journal default header borders*/

      borderwidth=2 pt;

   class header /

      just=left;

   class NoteContent /

      font=fonts("EmphasisFont");

   class systemtitle /

      font_size=16 pt

      just=center;

   class systemfooter /

      font_size=10 pt

      just=left

      textindent=1.4 in;
end;

run;

** Create Macro Variables to be used in Report Creation; ❷

%let title1   = 2011 Sales;

%let rndfoot  = Note: Sales are rounded up to the nearest dollar.;

%let dreffoot = Figure 1.2: Summary Report;

%let sreffoot = Figure 1.1: Detail Report;
```

(handwritten annotation: → Desired font for COMPUTE BLOCK of Summary block)

10 *PROC REPORT by Example: Techniques for Building Professional Reports Using SAS*

```
%let odsopt    = nodate nonumber orientation=portrait missing=" ";
%let outpath   = C:\Users\User\My Documents\SAS\BOOK\Programs\;
%let template  = JOURNALR;

** Set ODS Escape Character;  ❸
ods escapechar="^";

** Develop Macro to Modify Incoming Data for Reports;
%macro preproc;
   /** Concatenate data sets and variables **/
   data sales(drop=storenam);
      length type $20 storenm $40;
      set store(in=instore)  ❹
          ecomm(in=inecomm);

      /** Create Store Type Flags **/
      if inecomm then TYPE = "E-Commerce";  ❺
      else if instore then TYPE = "In-Store";

      /** Concatenate Store Number with Store Name **/
      storenm = catx(" / ",put(storenum,3.),storenam);  ❻
   run;
%mend preproc;
```

Ch1Setup.SAS creates a common ODS Style Template

❶ The Detail and Summary reports use a common style template. As a start, the SAS-supplied ODS style template "Journal" is used as the parent style. We name the modified template "JournalR" for this example.

Next, selected style attributes for the style elements Fonts, Table, Header, NoteContent, Systemtitle and Systemfooter are changed via CLASS statements.

- We assign desired font characteristics such as font face, font size, and font weight to various font names ("DocFont", "EmphasisFont", and "HeadingFont") in the CLASS FONTS statement. For example, we change the default Journal Style "HeadingFont", which applies to the column headers, to "Georgia", 16 pt., Bold Italic. We change "DocFont", which applies to data in table cells, to "Georgia", 9 pt.

- ○ Note, there is no comma between the font style of "Italic" and the font weight of "Bold," as in the HeadingFont specification.
- ○ We can later reference one of these sets of font characteristics by their corresponding font name when we want these attributes applied to a specific part of the output. For example, "EmphasisFont" is later applied in the CLASS NoteContent statement, informing SAS that this list of font characteristics should be applied to the NoteContent portion of the output (explained further below).

- The CLASS TABLE statement is used to modify the borders of the PROC REPORT table cells. For example, the "rules=none" specification in the Table section removes Journal's default header borders so that we may apply customized borders in the PROC REPORT code. We thicken the borders to be applied to a 2 pt. width.
- In the CLASS HEADER statement, we left justify the headers. (column headers)
- The "NoteContent" modifications allow us to obtain the desired font for the COMPUTE block lines and summary rows. We change the NoteContent font from the default to our "Emphasis Font" so the row headers (e.g., "Store" and "Region") will stand out more in the report.

Figure 1.3 displays the default Journal Style NoteContent font (See "Store" and "Region" lines), which is Courier 10 pt. In contrast, Figure 1.4 shows the report with our user-specified "EmphasisFont", which we chose as ("Georgia", 9 pt., Bold).

Figure 1.3 Default NoteContent ("Store" and "Region" text)

Store	
1 / Smith-Fine Goods	$991,322
2 / J.B. Prog & Co.	$744,388
3 / XYZ Warehouse	$345,675
4 / E-Home Store	$385,720
Region	
East	$926,031
West	$756,517
North	$385,438
South	$399,119

Figure 1.4 NoteContent ("Store" and "Region" text) after our specified EmphasisFont is applied

Store	
1 / Smith-Fine Goods	$991,322
2 / J.B. Prog & Co.	$744,388
3 / XYZ Warehouse	$345,675
4 / E-Home Store	$385,720
Region	
East	$926,031
West	$756,517
North	$385,438
South	$399,119

The CLASS SYSTEMTITLE and CLASS SYSTEMFOOTER statements are also used to provide template specifications that apply to the Detail and Summary reports. For example, the same Title and Footnote font size and horizontal justification applies to both reports. We provide the instructions for these title and footnote attributes in PROC TEMPLATE so we do not need to specify these in each later title and footnote statement.

For example, The PROC TEMPLATE code

```
class systemtitle /
  font_size=16 pt
  just=center;
```

achieves the same result as

```
title1 h=16pt j=center "Title goes here";
```

Likewise, the PROC TEMPLATE code

```
class systemfooter /
  font_size=10 pt
  just=left
  textindent=1.4 in;
```

achieves the same result as

```
footnote1 h=10pt j=left "^S={indent=1.4 in}Footnote goes here";
```

Recommended books for more information on PROC TEMPLATE and ODS Style Templates can be found in the references section of this book, including (Haworth, Zender, and Burlew (2009) and Smith (2013).

In addition to modifying the ODS Style Template, Ch1Setup.SAS creates macro variables, declares the ODS Escape Character, and prepares the input data set for the Detail and Summary reports.

❷   **Ch1Setup.SAS creates macro variables for:**
- Common titles – There is one common title, "2011 Sales", which has been assigned to &TITLE1.
- Common footnotes – There is one common footnote, "Sales are rounded up to the nearest dollar.", which has been assigned to &RNDFOOT.
- Additional footnotes – The Detail report will reference &DREFFOOT and the Summary Report will reference &SREFFOOT, which contain Figure numbers. These references are placed in the Setup program so any later changes to Figure numbers can be made globally from one location.
- Common options – The SAS system options we want applied to both ODS RTF reports are stored in &ODSOPT.
- Common output path – We want both reports to be stored in "C:\Users\User\My Documents\SAS\BOOK\Programs\", which has been assigned to &OUTPATH.
- Common template – Both reports will use the style template, JOURNALR, which has been assigned to &TEMPLATE.

❸   We tell ODS to use the caret character ("^") as the ODS ESCAPECHAR character value for in-line formatting.

Ch1Setup.SAS also creates the macro PREPROC to handle common data processing:

❹   Concatenate the source data sets STORE and ECOMM
❺   Assign store type variable (based on each source data set – ECOMM or STORE)
❻   Concatenate store names to store numbers

Table 1.3 displays a PROC PRINT of the new data set after calling the PREPROC macro. Note that the data set now contains both In-Store and E-Commerce data, designated by the variable TYPE. STORENM contains each store's number before the store name.

Table 1.3 PROC PRINT of Data SALES – Result of Macro PREPROC

Type	storenm	STORENUM	REG	SALES
In-Store	1 / Smith-Fine Goods	1	East	$200,310
In-Store	1 / Smith-Fine Goods	1	West	$189,356
In-Store	2 / J.B. Prog & Co.	2	East	$89,499
In-Store	2 / J.B. Prog & Co.	2	West	$44,250
In-Store	1 / Smith-Fine Goods	1	North	$99,346
In-Store	1 / Smith-Fine Goods	1	South	$107,220
In-Store	2 / J.B. Prog & Co.	2	North	$100,171
In-Store	2 / J.B. Prog & Co.	2	South	$79,465
In-Store	3 / XYZ Warehouse	3	East	$49,126
E-Commerce	1 / Smith-Fine Goods	1	East	$107,024
E-Commerce	1 / Smith-Fine Goods	1	West	$90,033
E-Commerce	2 / J.B. Prog & Co.	2	East	$120,050
E-Commerce	2 / J.B. Prog & Co.	2	West	$110,631
E-Commerce	1 / Smith-Fine Goods	1	North	$88,024
E-Commerce	1 / Smith-Fine Goods	1	South	$110,009
E-Commerce	2 / J.B. Prog & Co.	2	North	$97,897
E-Commerce	2 / J.B. Prog & Co.	2	South	$102,425
E-Commerce	3 / XYZ Warehouse	3	East	$159,902
E-Commerce	3 / XYZ Warehouse	3	West	$136,647
E-Commerce	4 / E-Home Store	4	East	$200,120
E-Commerce	4 / E-Home Store	4	West	$185,600

Writing the Detail Report Program

After running the Ch1Setup.SAS program and PREPROC macro for common report features, the only additional preparation needed prior to running PROC REPORT is to create a numeric version of the Region variable to be used in PROC REPORT for ordering rows. The numeric variable will be used in PROC REPORT for arranging the region rows in the order of "East," "West," "North," and "South."

Detail Report Pre-Processing Code

```
** %Include Ch1Setup.SAS and Run Pre-processing Macro to Create SALES
Data;

** Create Template and macro variables;

%include"C:\Users\User\Desktop\APR FINAL CH1_5\Programs\Ch1Setup.sas";

** Concatenate and create new data;

%preproc;

** Create Region Informat;

proc format;
   invalue region
      "East"  = 1
      "West"  = 2
      "North" = 3
      "South" = 4;
run;

** Create Region Order Variable;

data saledetr;
   set sales;
   regn=input(reg,region.);
run;
```

Detail Report: Titles, Footnotes, and ODS RTF Preparation

The Detail report data is now in the desired format for PROC REPORT. The next step is to add the desired titles and footnotes and prepare the ODS RTF destination.

Code for Titles, Footnotes, and ODS RTF Preparation

```
** Set Page Titles and Footnotes;   ❶
title1 "&title1 - Detail";
footnote1 "&rndfoot";
footnote2 "Reference: &dreffoot";

** ODS RTF Preparation;
ods _all_ close ;   ❷
options &odsopt;   ❸
ods rtf style=&template file="&outpath.detail.rtf" bodytitle;   ❹

**  {PROC REPORT CODE GOES HERE…};

ods rtf close;   ❺
ods html;   ❻
title;   ❼
footnote;   ❽
```

❶ title1, footnote1, and footnote2 reference previously created macro variables (&TITLE1, &RNDFOOT, and &DREFFOOT) to apply title and footnote text. The macro variables need to be placed within double quotes so they can be resolved.

The title appears in the RTF output as "2011 Sales - Detail." The footnotes are printed as "Note: Sales are rounded up to the nearest dollar." and "Reference: Figure 1.2: Summary Report."

❷ While closing other destinations is not necessary, the PROC REPORT output is needed only in the RTF destination; therefore the other destinations are closed. If a destination is closed, it is important to remember to reopen any desired destinations once you are ready to print to that destination again, otherwise, you may be wondering why your output is not created.

❸ Reference the ODS options specified in the Ch1Setup.SAS program.

❹ Open the RTF destination.
- Reference the style template specified in the Ch1Setup.sas program (&TEMPLATE)
- Reference the path specified in the Ch1Setup.sas program (&OUTPATH)

- Assign the specific file name for the detail report ("detail")
- These examples use the BODYTITLE option to place titles and footnotes directly above and below the table rather than in the MS Word headers and footers. With the BODYTITLE option, a title appears once before the top of the table and footnotes appear once after the end of the table. This works for these tables because an entire table fits on one page.
    - There are methods to place titles and footers directly above and below the table on every page. One method that uses BODYTITLE is shown at: http://support.sas.com/kb/36/288.html ("Sample 36288: Repeating text on an RTF page when the BODYTITLE option is in effect")
    - Another method which does not employ BODYTITLE increases the page header margins (with the HEADERY option) and footer margins (with the FOOTERY option) so that the titles and footnotes surround the table.

❺ Close the RTF destination

❻ Re-open any desired printing destinations, in this example HTML

❼, ❽ Reset the titles and footnotes if you don't want the most recent titles and footnotes carried through to later output

Producing the Report with PROC REPORT

Now that the ODS RTF destination has been opened, we are ready to produce the report. The PROC REPORT code below translates our SALEDETR data (see Table 1.4) into our PROC REPORT output (see Figure 1.5). Note that the data set SALEDETR does not need to be sorted prior to loading it into PROC REPORT. PROC REPORT options will be used to order rows.

Table 1.4 Partial PROC PRINT of SALEDETR Data

type	storenm	STORENUM	REG	SALES	regn
In-Store	1 / Smith-Fine Goods	1	East	$200,310	1
In-Store	1 / Smith-Fine Goods	1	West	$189,356	2
In-Store	2 / J.B. Prog & Co.	2	East	$89,499	1
In-Store	2 / J.B. Prog & Co.	2	West	$44,250	2
In-Store	1 / Smith-Fine Goods	1	North	$99,346	3
In-Store	1 / Smith-Fine Goods	1	South	$107,220	4
In-Store	2 / J.B. Prog & Co.	2	North	$100,171	3
In-Store	2 / J.B. Prog & Co.	2	South	$79,465	4

type	storenm	STORENUM	REG	SALES	regn
In-Store	3 / XYZ Warehouse	3	East	$49,126	1
E-Commerce	1 / Smith-Fine Goods	1	East	$107,024	1
E-Commerce	1 / Smith-Fine Goods	1	West	$90,033	2
E-Commerce	2 / J.B. Prog & Co.	2	East	$120,050	1
E-Commerce	2 / J.B. Prog & Co.	2	West	$110,631	2
E-Commerce	1 / Smith-Fine Goods	1	North	$88,024	3
E-Commerce	1 / Smith-Fine Goods	1	South	$110,009	4

Figure 1.5 Partial PROC REPORT Output

2011 Sales - Detail

Store	*Region*	*Store Type*	*Sales*
1 / Smith-Fine Goods	East	E-Commerce In-Store	$107,024 $200,310 **$307,334**
	West	E-Commerce In-Store	$90,033 $189,356 **$279,389**
	North	E-Commerce In-Store	$88,024 $99,346 **$187,370**
	South	E-Commerce In-Store	$110,009 $107,220 **$217,229**
2 / J.B. Prog & Co.	East	E-Commerce In-Store	$120,050 $89,499 **$209,549**

The following items are accomplished within the REPORT procedure:

- Specify Variables to be Reported, Left to Right in Desired Order with COLUMN Statement.
- Define Variable Usage, Labels, and Style Elements with DEFINE Statements.
- Summarize Sales after each Region.
- Insert a Blank Line after Each Region.

Detail Report - PROC REPORT Code

```
** Detail Report;

** Style Report, Column Data and Headers, and Lines;  ❶
proc report data = saledetr nowd center split="|"
  style(report)=[cellpadding=0 pt]
  style(header)=[vjust=middle borderbottomwidth=2 pt cellheight=.8 in];

** Specify Variables to be Reported, in Desired Order;  ❷
  column STORENM REGN REG TYPE SALES;

** Define Variable Usage, Labels, and Style Elements;  ❸
  define STORENM / order "Store" style(column)=[cellwidth=1.4 in
                              font_weight=bold];

  define REGN    / noprint order order=internal;


  define REG    / order "Region" style(column)=[cellwidth=1.4 in
                              indent=.55 in]
                              style(header)=[indent=.25 in];

  define TYPE   / order "Store Type" style(column)=[cellwidth=1.4 in
                              indent=.3 in]
                              style(header)=[indent=.1 in];

  define SALES / "Sales" style(column)=[cellwidth=1 in just=dec]
```

```
                           style(header)=[just=r];

** Summarize Sales after each Region;  ❹
   break after REG / summarize suppress
                     style(summary)=[font_weight=bold
                                     foreground=darkblue];

** Insert a Blank Line Before each Region;  ❺
   compute before REG;
      line " ";
   endcomp;
run;
```

Key PROC REPORT statements and options used in this example:

❶ Overrides to the STYLE template are applied in the PROC REPORT statement for style elements specific to the Detail report. The overrides are optional and presented primarily for the purpose of demonstrating style overrides to the reader, since they are used throughout the book.

- Cellpadding, the amount of space between the table cell border and its content is set to 0 pt with the STYLE(REPORT)= option. This helps to prevent wrapping, and to decrease the amount of white space within the report so that the full Detail report can fit on one page.
- The header elements and attributes are also modified with the STYLE(HEADER)= option in order to make the report more legible. The headers are centered vertically with the specification VJUST=MIDDLE. A bottom border is created for the headers and is set to a width of 2 pt. The header cell height is increased to .8 inches.

Without the overrides specified in the PROC REPORT statement the Detail report would display as shown in Figure 1.6. In contrast, Figure 1.7 shows the Detail report with the REPORT and HEADER style overrides.

Figure 1.6: Detail Report WITHOUT the STYLE(REPORT)= and STYLE(HEADER)= Overrides

2011 Sales - Detail

Store	*Region*	*Store Type*	*Sales*
1 / Smith-Fine Goods	East	E-Commerce	$107,024
		In-Store	$200,310
			$307,334
	West	E-Commerce	$90,033
		In-Store	$189,356
			$279,389

Figure 1.7: Detail Report WITH the STYLE(REPORT)= and STYLE(HEADER)= Overrides

2011 Sales - Detail

Store	*Region*	*Store Type*	*Sales*
1 / Smith-Fine Goods	East	E-Commerce	$107,024
		In-Store	$200,310
			$307,334
	West	E-Commerce	$90,033
		In-Store	$189,356
			$279,389

❷ The **COLUMN** statement specifies the columns to be included left to right. The COLUMN statement also determines the column order in the report, therefore columns are listed in the order in which they should be processed and/or appear in the report (STORENM, REGN, REG, TYPE, and then SALES). Without a COLUMN statement specification, the column order would reflect the physical order of the incoming data set.

❸ We specify variable usage, labels, and style elements in the **DEFINE** statements.

The **ORDER** options arrange rows in order of these columns' values respectively (STORENM, REGN, REG, TYPE). Without the COLUMN and ORDER specifications the rows would not be ordered. SAS would assign the character variables their default usage type, which is DISPLAY, and rows would reflect the physical order of the incoming data set.

- The ORDER usage option without the ORDER= option (e.g., see DEFINE statements for STORENM, REG, TYPE) assumes the default ORDER= option, which is FORMATTED at the time of writing this book. ORDER=FORMATTED orders rows in the order of formatted values, in this case alphanumeric order.
- The ORDER usage option with the ORDER=INTERNAL specification (e.g., REGN) orders rows in the order of unformatted values. Note that in the report, regions are ordered by the unformatted value (1=East, 2= West, 3=North, 4=South) versus alphabetical order. REGN is used for the sole purpose of ordering rows. We suppress the printing of this column with the NOPRINT option.

The ORDER option also:

- Prevents repetition of values over multiple rows (e.g. we want the text "1 / Smith-Fine Goods" to appear only on the first occurrence of this value).
- Allows for use of the BREAK AFTER VARNAME and COMPUTE BEFORE VARNAME before/after each new REG value. (BREAK and COMPUTE based on a BEFORE/AFTER location require GROUP or ORDER usage for that variable).

We apply **style overrides** in some of the DEFINE statements to override the style template's attributes for specific columns and headers. This is done with the STYLE(COLUMN)= and STYLE(HEADER)= options.

- For example, for the variable STORENM, we set the column cell width to 1.4 inches. In addition we apply bold font weight to the list of Store Names in the column.
- For the variable REG, we set the column cell width to 1.4 inches, and indent the data in the columns .55 inches. We indent the column header .25 inches.

❹ The sales totals within each Store-Region (summary lines) are generated by the SUMMARIZE option specified in the **BREAK** statement. Although the ANALYSIS usage option and the SUM statistic are not explicitly specified in the DEFINE statement for Sales, these are applied because ANALYSIS is the default usage for numeric variables, and the SUM statistic is the default behavior for row summaries.

The style of these totals (other than the default italics) are specified in the PROC REPORT statement with **STYLE(SUMMARY)**=[font_weight=bold foreground=darkblue]

❺ The **COMPUTE** block creates a blank row before each new value of region (REG) within each store (STORENM).

Writing the Summary Report Program

The Summary Report requires more pre-processing than did the Detail Report to get the data into the desired format for PROC REPORT.

The key Summary Report pre-processing steps include:

- Map separate variables/values to one column for PROC REPORT (CATNAME for "Category" column)
- Derive numeric variables that will be used to assign report row order

Map Separate Variables/Values to One Column for PROC REPORT

Note that in the PROC REPORT output the originally "separate" variable names, STORENAM (Store), REG (Region), and TYPE (Store Type) and their corresponding values appear in one column, "Category."

Figure 1.8 Originally Separate Variables are Stacked in Category Column

2011 Sales - Summary

Category	Sales
Total	**$2,467,105**
Store	
1 / Smith-Fine Goods	$991,322
2 / J.B. Prog & Co.	$744,388
3 / XYZ Warehouse	$345,675
4 / E-Home Store	$385,720
Region	
East	$926,031
West	$756,517
North	$385,438
South	$399,119
Store Type	
E-Commerce	$1,508,362
In-Store	$958,743

Note: Sales are rounded up to the nearest dollar.
Reference: Figure 1.1: Detail Report

There are a number of ways to restructure the data in this format. PROC TRANSPOSE is used in this example to map all of the category names and values to a new variable named CATNAME.

Summary Report - Pre-Processing Code

```
** %Include Ch1Setup.SAS and Run Pre-processing Macro to Create SALES
Data;

/** Create Template and macro variables **/

%include"C:\Users\User\Desktop\APR FINAL CH1_5\Programs\Ch1Setup.sas";


/** Concatenate and create new data **/
```

```
%preproc;

** Create TOTAL variable for Total Row in PROC REPORT;
** All records are set to same value ("Total") for later sum;
data salesumr;
   set sales;
   total = "Total";
run ;

** Transpose data to get needed items mapped to Category (CATNAME);
proc sort data=salesumr;
   by sales storenum;
run;

proc transpose data=salesumr out=salesum2(rename=(col1=catname));
   by sales storenum;
   var total storenm reg type;
run;
```

Figure 1.9 displays a partial PROC PRINT of the transposed data, sorted by _NAME_ and STORENUM. A sort is not needed but is performed so you can better see how the sales data will eventually translate into sums. The variables that you want to end up in one column (COL1, renamed to CATNAME) are specified in the PROC TRANSPOSE VAR statement. In the partial PROC PRINT (Figure 1.9) you can see that _NAME_ contains our original variable names, for example STORENM. CATNAME contains the values for these items.

Figure 1.9 Partial PROC PRINT of Transposed Data

NAME	STORENUM	catname	SALES
storenm	1	1 / Smith-Fine Goods	$88,024
storenm	1	1 / Smith-Fine Goods	$90,033
storenm	1	1 / Smith-Fine Goods	$99,346
storenm	1	1 / Smith-Fine Goods	$107,024
storenm	1	1 / Smith-Fine Goods	$107,220
storenm	1	1 / Smith-Fine Goods	$110,009
storenm	1	1 / Smith-Fine Goods	$189,356
storenm	1	1 / Smith-Fine Goods	$200,310
storenm	2	2 / J.B. Prog & Co.	$44,250
storenm	2	2 / J.B. Prog & Co.	$79,465
storenm	2	2 / J.B. Prog & Co.	$89,499
storenm	2	2 / J.B. Prog & Co.	$97,897
storenm	2	2 / J.B. Prog & Co.	$100,171
storenm	2	2 / J.B. Prog & Co.	$102,425
storenm	2	2 / J.B. Prog & Co.	$110,631
storenm	2	2 / J.B. Prog & Co.	$120,050

$991,322 ❶ (sum for store 1)

$744,388 ❷ (sum for store 2)

The transposed data rows above are not yet in the desired order for our Summary Report, but you can get the idea that the data is now structured so the observations within groups can be summed, as in Figure 1.10.

Figure 1.10 Sale Summed in Summary Report

2011 Sales - Summary

Category	*Sales*
Total	$2,467,105
Store	
1 / Smith-Fine Goods	❶ $991,322
2 / J.B. Prog & Co.	❷ $744,388
3 / XYZ Warehouse	$345,675
4 / E-Home Store	$385,720

Assign Report Order to Variables

Two numeric variables, CATORD and SUBCTORD, are created specifically for ordering data later on in the REPORT procedure.

CATORD will be used to order categories and to insert blank rows after each group. Total is the first category (given a value of 0). Store and its associated values (1 / Smith-Fine Goods, 2 / J.B. Prog, etc.) are the second category (given a value of 1). Region and its associated values are the third category (given a value of 2), and so on.

SUBCTORD will be used to order rows **within** each category. Within Store, stores will be sorted by Store Number (1, 2, 3, and 4). Within Region, East will be shown first, then West, etc.

Table 1.5 shows the mapping of rows to CATORD and SUBCTORD which will translate into row order in the Summary Report.

Table 1.5 Derive CATORD and SUBCTORD for Ordering Rows

catname	catord	subctord
Total	0	0
1 / Smith-Fine Goods	1	1
2 / J.B. Prog & Co.	1	2
3 / XYZ Warehouse	1	3
4 / E-Home Store	1	4
East	2	1002
West	2	1003
North	2	1004
South	2	1005
E-Commerce	3	1007
In-Store	3	1008

Code for Creating Ordered Variables

```
** Informats for Summary Report; ❶
proc format;
   invalue catord
      "TOTAL"     = 0
      "STORENM"   = 1
      "REG"       = 2
      "TYPE"      = 3;
   invalue rowlabl
      "Total"     = 0
      "Region"    = 1
      "East"      = 2
      "West"      = 3
```

```
           "North"       = 4
           "South"       = 5
           "Store Type"  = 6
           "E-Commerce"  = 7
           "In-Store"    = 8;
run;

data salesum2;
  set salesum2;
  ** Create CATORD for ordering groups of rows in PROC REPORT; ❷
    catord = input(upcase(_name_),catord.);
  ** Create SUBCTORD to order sub-category rows in PROC REPORT;
    if catord = 0 then subctord = 0; ❸
    else if catord = 1 then subctord = storenum; ❹
    else subctord = input(catname,rowlabl.) + 1000; ❺
run;
```

❶ Informats are created to reflect the desired row reporting order of categories and subcategories.

❷ A CATORD value is assigned for every row.

❸ Note that the "Total" rows (where CATORD=0) are assigned a SUBCTORD value of 0 to ensure "Total" is reported before the first Store Number (1).

❹ For Store (names) the assumption is that the list of stores will grow, and in future runs of our program we don't want to have to add new formats to our list each time new stores are added or a store number changes. Therefore, rather than using informats to assign stores a number, an assignment statement that captures store number is used.

❺ For the shorter and/or finite lists for Region and Store Type, an informat is used to assign SUBCTORD. Note the "+ 1000" added to the Region and Store Type values (because these variables occur after STORE) to ensure these rows are ordered after the last store number, which we expect to increase over time.

Summary Report: Titles, Footnotes, and ODS RTF Preparation

The data set is now in the needed structure for PROC REPORT. The next step is to prepare titles, footnotes, and ODS RTF setup.

Code for Titles, Footnotes, and ODS RTF Preparation

```
** Titles and Footnotes;
title1    "&title1 - Summary";
footnote1 "^S={indent=2 in}&rndfoot";
footnote2 "^S={indent=2 in}Reference: &sreffoot";

** ODS RTF Setup;
ods _all_ close ;
options &odsopt;
ods rtf style=&template file="&outpath.salessum.rtf" bodytitle;

** {PROC REPORT CODE GOES HERE…};
ods rtf close;
ods html;
title;
footnote;
```

The process for applying titles and footnotes is the same as that for the Detail Report section. We're simply changing some of the titles and footnotes that apply to the Summary report. Beside the verbiage, we change the indentation of the Summary report footnote (from 1.4 inches to 2 inches) to align with this smaller table. The style override is applied by inserting the inline formatting function, ^S={indent=2 in} within the quotation marks around each footnote. The title appears in the RTF output as "2011 Sales – Summary". The footnotes appear as "Note: Sales are rounded up to the nearest dollar." and "Reference: Figure 1.1: Detail Report".

The ODS RTF statements are the same as for the Detail report section with the exception of the file name (salessum) specification.

Producing the Report with PROC REPORT

Now that our data is in the desired structure, a small amount of PROC REPORT code will translate our SALESUM2 data (Table 1.6) into our Summary Report (Figure 1.11).

Table 1.6 Partial PROC PRINT of Final Sales Summary Data

catname	catord	subctord	SALES
Total	0	0	$44,250
Total	0	0	$49,126
Total	0	0	$79,465
Total	0	0	$88,024
Total	0	0	$89,499
Total	0	0	$90,033
Total	0	0	$97,897
Total	0	0	$99,346
Total	0	0	$100,171
Total	0	0	$102,425
Total	0	0	$107,024
Total	0	0	$107,220
Total	0	0	$110,009
Total	0	0	$110,631
Total	0	0	$120,050
Total	0	0	$136,647
Total	0	0	$159,902
Total	0	0	$185,600
Total	0	0	$189,356
Total	0	0	$200,120
Total	0	0	$200,310
1 / Smith-Fine Goods	1	1	$88,024
1 / Smith-Fine Goods	1	1	$90,033
1 / Smith-Fine Goods	1	1	$99,346
1 / Smith-Fine Goods	1	1	$107,024

catname	catord	subctord	SALES
1 / Smith-Fine Goods	1	1	$107,220
1 / Smith-Fine Goods	1	1	$110,009
1 / Smith-Fine Goods	1	1	$189,356
1 / Smith-Fine Goods	1	1	$200,310
2 / J.B. Prog & Co.	1	2	$44,250
2 / J.B. Prog & Co.	1	2	$79,465
2 / J.B. Prog & Co.	1	2	$89,499
2 / J.B. Prog & Co.	1	2	$97,897
2 / J.B. Prog & Co.	1	2	$100,171
2 / J.B. Prog & Co.	1	2	$102,425
2 / J.B. Prog & Co.	1	2	$110,631
2 / J.B. Prog & Co.	1	2	$120,050
3 / XYZ Warehouse	1	3	$49,126
3 / XYZ Warehouse	1	3	$136,647
3 / XYZ Warehouse	1	3	$159,902

Figure 1.11 Transformation of PROC PRINT to PROC REPORT Output

2011 Sales - Summary

Category	*Sales*
Total	**$2,467,105**
Store	
1 / Smith-Fine Goods	$991,322
2 / J.B. Prog & Co.	$744,388
3 / XYZ Warehouse	$345,675
4 / E-Home Store	$385,720
Region	
East	$926,031
West	$756,517
North	$385,438
South	$399,119
Store Type	
E-Commerce	$1,508,362
In-Store	$958,743

Note: Sales are rounded up to the nearest dollar.
Reference: Figure 1.1: Detail Report

Summary Report - PROC REPORT Code

```
{Add to the PROC FORMAT section...} ❶
value rowheadr
   0 = " "
   1 = "Store"
   2 = "Region"
   3 = "Store Type";

** Summary Report;
proc report data = salesum2 nowd missing center split="|"
             style(header)=[vjust=bottom cellheight=.55 in]
             style(lines) =[vjust=top just=l font_size=10 pt]
             OUT=PRSSUM; ❷
   column catord subctord catname sales;
   define catord   / group order=internal noprint; ❸
   define subctord / group order=internal noprint; ❸
   define catname  / group order=internal "Category" ❸
                     style(column)=[cellwidth=2.4 in];
   define sales    / "Sales" style(column)=[cellwidth=1.6 in just=r]
                     style(header)=[just=r];

** Change Total Row Style Attributes and Indent Subcategory Row
Headers; ❹
   compute catname /char length=30;
     if catname ne "Total" then
        call define("catname","style","style={indent=.25 in}");
     else
        do;
```

```
            call define(_row_,"style","style={bordertopwidth=2 pt
                                    cellheight=.5 in vjust=middle
                                    font_size=10 pt
                                    font_weight=bold}");
      end;
   endcomp;

   ** Print Category Row Headers; ❺
   compute before catord;
      line catord rowheadr.;
   endcomp;
run;
```

PROC REPORT will be able to translate our Table 1.6 data structure into Figure 1.11 by…

❶ Adding a format to assign the values we would like to see in the report for the main row headers "Store","Region","Store Type"). The "Total" row does not get a row header.

❷ Applying style attributes to the column headers and COMPUTE block lines.
- The column header style override (accomplished with STYLE(HEADER=) option) adjusts the vertical justification, placing the header text at the bottom of the cell. The cell height is reset to .55 inches.
- The COMPUTE block line statement style override (accomplished with STYLE(LINES=) option) resets the vertical justification of the Row headers to be at the top of the cell, and horizontal justification=to be left. The font size is set to 10 pt. Note that without the JUST=L style override for LINES, the row headers would be centered as shown in Figure 1.12.

Figure 1.12 - Summary Report Without JUST = L Style Override for LINES

2011 Sales - Summary

Category	*Sales*
Total	**$2,467,105**
Store	
1 / Smith-Fine Goods	$991,322
2 / J.B. Prog & Co.	$744,388
3 / XYZ Warehouse	$345,675
4 / E-Home Store	$385,720
Region	
East	$926,031
West	$756,517
North	$385,438
South	$399,119
Store Type	
E-Commerce	$1,508,362
In-Store	$958,743

❸ We consolidate rows and sum the ANALYSIS variable SALES by applying the GROUP option to CATORD, SUBCTORD, and CATNAME. In this context, "Consolidate" means to collapse a group of rows into a summary row. For example, the store "1 / Smith-Fine Goods" gets one row showing its summed sales. The store "2 / J.B. Prog & Co." gets one row showing its summed sales, and so on.

While the primary reason for using the GROUP option here is to consolidate rows, the option also orders the rows for us. The **COLUMN** statement's specified order of columns (CATORD SUBCTORD CATNAME) along with the **GROUP** option specified in the DEFINE statements arrange rows in order of these variables' values respectively. The **GROUP** option also allows for use of the COMPUTE before each new CATORD value.

❹ We use a COMPUTE block to indent the rows showing levels of variables, and leave the variable names (shown as row headers) un-indented.

CALL DEFINE statements are used to conditionally apply style attributes to the "Total" row versus the row headers. Note that CALL DEFINE can be used only in a COMPUTE block that is attached to a report item. In this case, our report item CATNAME is already part of the input data set. We COMPUTE the variable CATNAME so we can apply the CALL DEFINEs, but we still define CATNAME's usage as GROUP so we can consolidate rows.

- In order to make the Total sales row stand out, we increase the font size, font weight (bold), and cell height where CATNAME equals "Total". A top border is added and the cell text is centered vertically ("VJUST=MIDDLE"). Because we want these style attributes applied to the entire "Total" row, we identify _ROW_ (without quotation marks) as the item to which the CALL DEFINE attributes should be applied, as in "call define(_row_,........)".
- The subcategory row headers (e.g., "East," "West," "North," and "South") are indented .25 inches. (The subcategories occur where CATNAME does not equal "Total".) In the case of the subcategory rows we only apply the indentation to the CATNAME column. Therefore the CALL DEFINE is applied only to the column "CATNAME". The column must be enclosed in quotation marks, as in call define("catname", ...)".

❺ A **COMPUTE** block LINE statement is used to print the appropriate row header text ("Store", "Region", and "Store Type") before each new value of CATORD.

For the reader's reference, the output data set created by the PROC REPORT code is shown in Table 1.7. The optional output dataset, named PRSSUM here, is created with the OUT= option in the PROC REPORT statement.

It is worth noting that the COMPUTE BEFORE results are represented differently in the output data set (Table 1.8) than in the printed report (Figure 1.13). Table 1.7 notes some of these differences.

Table 1.7 COMPUTE BEFORE Representation in Output Data Set Versus Printed Report

Output Data Set PRSSUM (Table 1.8)	Printed Report (Figure 1.13)
The COMPUTE block LINE statement results do not appear in the output data set.	The COMPUTE block LINE statements display as row headers, "Store," "Region," and "Store Type" corresponding to the CATORD values 1 through 3 respectively.
COMPUTE BEFORE CATORD generates a summary line BEFORE each distinct value of catord, in which the analysis variable SALES is summarized for each catord group in the data set.	If we want these summary sales rows to print in the printed report, we need a BREAK statement with a / SUMMARIZE option. Since this is not desired for the printed report, a customized summary line is created with a LINE statement.
The output data set contains an automatic internal variable named "_BREAK_." In this case, _BREAK_ is populated with the value "catord" due to the COMPUTE BEFORE CATORD statement.	This variable is not a report item and is not printed.
CATORD and SUBCTORD are still part of PROC REPORT's data set in spite of the NOPRINT option.	CATORD and SUBCTORD variables do not display in the printed report due to the NOPRINT option.

Table 1.8 PROC PRINT OF PRSSUM Data Set

catord	subctord	catname	SALES	_BREAK_
0			$2,467,105	catord
0	0	Total	$2,467,105	
1			$2,467,105	catord
1	1	1 / Smith-Fine Goods	$991,322	
1	2	2 / J.B. Prog & Co.	$744,388	
1	3	3 / XYZ Warehouse	$345,675	
1	4	4 / E-Home Store	$385,720	
2			$2,467,105	catord
2	1002	East	$926,031	
2	1003	West	$756,517	
2	1004	North	$385,438	
2	1005	South	$399,119	

catord	subctord	catname	SALES	_BREAK_
3			$2,467,105	catord
3	1007	E-Commerce	$1,508,362	
3	1008	In-Store	$958,743	

Figure 1.13 PROC REPORT Output

2011 Sales - Summary

Category	Sales
Total	**$2,467,105**
Store	
1 / Smith-Fine Goods	$991,322
2 / J.B. Prog & Co.	$744,388
3 / XYZ Warehouse	$345,675
4 / E-Home Store	$385,720
Region	
East	$926,031
West	$756,517
North	$385,438
South	$399,119
Store Type	
E-Commerce	$1,508,362
In-Store	$958,743

Note: Sales are rounded up to the nearest dollar.
Reference: Figure 1.1: Detail Report

Chapter 1 Summary

This chapter showed steps the programmer can implement that will allow for accuracy and ease of use to the reader when analyzing multiple reports. These include:

- A consistent report template
    - Example: Use of the same ODS Style template, with like style elements and attributes across reports
- Consistency of data definitions across reports
    - Example: Macro variables and macro used to merge E-Commerce and In-Store data, creation of the TYPE variable to designate these items, and concatenation of Store Name with Store Number
- Consistent labels across the reports
    - Example: titles, footnotes, headers, value labels (Store Names, the Regions, Store Type values)
- Footnote references clarifying which reports correspond to each other
    - Example: The Detail report's footnote "Reference: Figure 1.2: Summary Report" indicates that these two reports should be used in conjunction with one another. Likewise, the Summary report footnote references Figure 1.1: Detail Report.
- Check that numbers/statistics correspond across reports
    - Example: All of the "1 / Smith-Fine Goods" sales in the Detail report sum to $991,322, the total "1 / Smith-Fine Goods" sales reported in the Summary report.

Chapter 2: Formatting Highly Detailed Reports

Introduction 42

Example: Format National Sales Report 42

Goals for Formatting the National Sales Report 45

    Key Steps 45

Source Data 46

ODS Style Template Used 46

Programs Used 47

PROC TEMPLATE Program to Create New Style Template 47

The "Before Formatting" Program (Program 2.1) 49

Implementation 53

Transforming Figure 2.1 Into Figure 2.2 53

Displaying Region as a Line Above Each Report Page 53

    Overview of Region Display 53

    Code to Make the Region Display in Figure 2.3 54

Displaying Store and Branch Column Data in Bold Blue Font 54

    Code for Store and Branch Display 55

42 *PROC REPORT by Example: Techniques for Building Professional Reports Using SAS*

How to Insert Arrows for Quick Reference to Sales Increases/Decreases 56

    Overview on Arrow Insertion 56

    Code for Arrow Insertion.. 56

How to Add Spanning Headers, Bottom Cell Borders, and Underlines 59

    Highlights on Adding Spanning Headers, Borders, and Underlines 59

    Code for Adding Spanning Headers, Borders, and Underlines 60

Adding Blank Columns to Make the Report More Legible...................................... 62

    Overview of Adding Blank Columns 62

    Code for Adding Blank Columns... 62

Style: Add a Blank Line After Each Summary Line 65

    Highlights on Styling Summary Line and Adding a Blank Line................................... 65

    Code for Styling Summary Line and Adding a Blank Line 65

Chapter 2 Summary 66

Introduction

Consider a highly detailed report that is overwhelming with numbers, page after page. The results in this type of report can be very difficult for the reader to decipher. However, often, just slight changes in style and formatting can transform a report crammed with details into one in which the reader can identify important items and make sense of the results.

Example: Format National Sales Report

This example demonstrates how to modify a national sales report that displays quarterly sales by Region, Store Number, and Branch Number. Figures 2.1 and 2.2 show BEFORE and AFTER versions of the same report after implementing various ODS and PROC REPORT options. Almost all report enhancements are made using PROC REPORT along with ODS RTF enhancements. The only pre-processing of data in this example is to add a variable named BLANK that will be used to add empty columns in the REPORT procedure.

Figure 2.1 BEFORE Formatting

Region	Store #	Branch #	Quarter 4 2010	Quarter 1 2011	Quarter 2 2011
East	1001	1	$43,564	$42,555	$47,000
		2	$47,235	$47,523	$48,102
		3	$50,244	$45,999	$46,582
		4	$37,665	$37,778	$39,624
			$178,708	*$173,855*	*$181,308*
	1002	1	$50,000	$50,248	$49,335
		2	$48,045	$46,822	$48,000
		3	$60,023	$62,410	$62,650
		4	$55,000	$56,102	$56,109
		5	$52,345	$52,650	$54,197
		6	$56,444	$57,720	$57,800
		7	$59,721	$63,000	$65,254
			$381,578	*$388,952*	*$393,345*
	1003	1	$67,443	$66,230	$67,503
		2	$69,903	$70,085	$73,598
		3	$57,900	$60,000	$62,900
		4	$58,989	$59,222	$58,307
		5	$56,112	$57,364	$57,647
			$310,347	*$312,901*	*$319,955*
	1007	1	$51,008	$51,008	$52,310
		2	$46,733	$43,062	$45,564
		3	$61,228	$59,838	$62,300
			$158,969	*$153,908*	*$160,174*
	1008	1	$63,328	$63,000	$64,274
		2	$61,377	$66,461	$70,468
		3	$59,821	$59,637	$61,272
			$184,526	*$189,098*	*$196,014*
	1009	1	$60,044	$61,900	$64,528
			$60,044	*$61,900*	*$64,528*
	1011	1	$47,889	$42,500	$40,389
			$47,889	*$42,500*	*$40,389*

Figure 2.2 AFTER Formatting

Region: East

			2011 Sales			
		2010 Sales Quarter 4	**Quarter 1**	**Vs. Q4**	**Quarter 2**	**Vs. Q1**
<u>Store #</u>	<u>Branch #</u>					
1001	1	$43,564	$42,555	↓	$47,000	↑
	2	$47,235	$47,523	↑	$48,102	↑
	3	$50,244	$45,999	↓	$46,582	↑
	4	$37,665	$37,778	↑	$39,624	↑
		$178,708	*$173,855*	↓	*$181,308*	↑
1002	1	$50,000	$50,248	↑	$49,335	↓
	2	$48,045	$46,822	↓	$48,000	↑
	3	$60,023	$62,410	↑	$62,650	↑
	4	$55,000	$56,102	↑	$56,109	↑
	5	$52,345	$52,650	↑	$54,197	↑
	6	$56,444	$57,720	↑	$57,800	↑
	7	$59,721	$63,000	↑	$65,254	↑
		$381,578	*$388,952*	↑	*$393,345*	↑
1003	1	$67,443	$66,230	↓	$67,503	↑
	2	$69,903	$70,085	↑	$73,598	↑
	3	$57,900	$60,000	↑	$62,900	↑
	4	$58,989	$59,222	↑	$58,307	↓
	5	$56,112	$57,364	↑	$57,647	↑
		$310,347	*$312,901*	↑	*$319,955*	↑
1007	1	$51,008	$51,008	○	$52,310	↑
	2	$46,733	$43,062	↓	$45,564	↑
	3	$61,228	$59,838	↓	$62,300	↑
		$158,969	*$153,908*	↓	*$160,174*	↑
1008	1	$63,328	$63,000	↓	$64,274	↑

The printed book is presented in Grayscale and does not display the actual colors applied to Figure 2.2 (AFTER Formatting). Blue font was applied to elucidate key details, such as location data (Region, Store #s and Branch #s) and total Store sales. Green font was applied to up arrows (increases) and red font was applied to down arrows (decreases) to further highlight the direction of each store's sales from the previous quarter. Visit the author's web page at http://support.sas.com/publishing/authors/fine.html to see a color version of Figure 2.2.

Goals for Formatting the National Sales Report

The goals for formatting highly detailed reports are those that make the report easier on the eye.

The goals for this chapter include:

- Adding white space between rows and columns.
- Making key items stand out with bold font, borders, underlines, and color.
- Inserting symbols, in this case arrows, for quick reference.

Key Steps

Figure 2.1, the "BEFORE" version, is transformed into Figure 2.2, the "AFTER" version, by making the following modifications:

1. Region is displayed as a line above each report page, rather than as a column on each page.
2. Store #s and Branch #s are displayed in bold blue font (font color or text color is called "foreground" in the SAS code).
3. Arrows are displayed after each 2011 Sales column to provide a quick reference to whether sales increased or decreased, and open circles are displayed to indicate that sales stayed the same from the previous quarter.
   - Color is applied to the arrows to further highlight increases (green), decreases (red), and no change (black).
4. A spanning header, "2011 Sales" is added and bottom borders and underlines are added to header cells.
5. Summary lines, which show total Store sales, are brought out by applying blue font and adding a blank line after each summary line, before the start of each new store number.

6. Blank columns are added after the following columns: "Branch #," "2010 Sales Quarter 4," and after "Vs Q4,"

Each modification is explained in a later section of this chapter.

Source Data

There is one source data set, Ch2Sales, which contains quarterly sales totals by Region, Branch Number, and Store Number. Table 2.1 displays the variable information for SALES. Table 2.2 contains a sample of the Ch2Sales data so the user can visualize the structure of the data set.

Table 2.1 Ch2Sales Variable Information

Variables in Creation Order

#	Variable	Type	Len	Label
1	REGION	Char	8	Region
2	STORE	Char	8	Store #
3	BRANCH	Num	8	Branch #
4	QTR4_2010	Num	8	Quarter 4 2010 Sales
5	QTR1	Num	8	Quarter 1 2011 Sales
6	QTR2	Num	8	Quarter 2 2011 Sales

Table 2.2 Sample Ch2Sales Data

REGION	STORE	BRANCH	QTR4_2010	QTR1	QTR2
East	1001	1	43564	42555	47000
East	1001	2	47235	47523	48102
East	1001	3	50244	45999	46582
East	1001	4	37665	37778	39624
East	1002	1	50000	50248	49335
East	1002	2	48045	46822	48000

ODS Style Template Used

The same ODS Style template, named "RSTYLERTF" was used to produce both Figure 2.1 and Figure 2.2 reports. This is a user-created template, meaning YOU need to create it with PROC TEMPLATE. Note that the template RSTYLERTF is later referenced in the ODS statement prior to running the REPORT procedure (see Program 2.1's ODS statement starting with, "**ods rtf style=RSTYLERTF...**" in the Programs Used section).

> To reproduce Figure 2.1 or Figure 2.2 you will need to to run the PROC TEMPLATE Program shown in the PROGRAMS USED section.

Programs Used

The PROC TEMPLATE program is run first to create the ODS style template, which provides the base formatting for both Figure 2.1 and Figure 2.2.

The "Before Formatting" program (referred to as Program 2.1 for the remainder of this chapter) was used to create Figure 2.1. Both the PROC TEMPLATE program and Progam 2.1 are displayed in this chapter.

The final program, called Ch2Format.sas, was used to create Figure 2.2. Visit the author's web page at http://support.sas.com/publishing/authors/fine.html to download a copy of this program.

PROC TEMPLATE Program to Create New Style Template

We create a new ODS style template because we want to slightly alter the default RTF ODS style template prior to producing the reports. We name the new template RSTYLERTF with the DEFINE STYLE statement. The PARENT= statement specifies that RSTYLERTF should inherit the style elements from STYLES.RTF.

We make the desired style element modifications for our new template via CLASS statements. Specifically, we remove some of the table and cell borders, thicken the table's top border, remove the background color from the header cell, and modify the page margins.

```
** PROC Template Program;
proc template;
  define style rstyleRTF;
    parent=styles.rtf;
  class table/
    rules=none
```

```
        frame=above
        borderwidth=1.5 pt
        cellpadding=0;
    class Header/
        background=_undef_;
    class body/
        bottommargin = .75 in
        topmargin    = 1.25 in
        rightmargin  = .75 in
        leftmargin   = .75 in;
    end;
    run;
```

> The RSTYLERTF template will inherit the style elements from the PARENT template (STYLES.RTF) except the table, header, and body elements specified by the CLASS statements.

Table 2.3 demonstrates some of the PROC TEMPLATE changes.

Table 2.3 PROC TEMPLATE Table and Header Changes

RTF DEFAULT ❶	→	TABLE Change: Rules=None ❷	→	TABLE Change: Frame=Above and Borderwidth=1.5 pt. ❸	→
Region		Region		Region	
East		East		East	
West		West		West	
North		North		North	
South		South		South	

TABLE Change:	→	HEADER Change:
Cellpadding=0 ❹		Background=_undef_ ❺

Region		Region
East		East
West		West
North		North
South		South

❶ This shows the default RTF ODS Style Template (the PARENT of the new ODS Style Template, RSTYLERTF).

❷ Rules=None requests that there are no border lines between table cells.

❸ Only the above portion of the table frame is requested. This above border is widened (made thicker) to 1.5 pt.

❹ The cell padding is set to 0 to remove the padding around the text within the cells. This has the effect of reducing the row heights.

❺ The background of the "Region" header is specified as undefined to remove the default background color.

The "Before Formatting" Program (Program 2.1)

Figure 2.1, the "BEFORE" figure was generated from the following program, after running the PROC TEMPLATE program.

```
** The "Before Formatting" Program (Program 2.1);
** Get Data;
data sales;
   set sasuser.ch2sales;
```

```
        run;

    ** Produce the Report;

        ods _all_ close;

        ods escapechar = "^";

        options nodate nonumber orientation=portrait;
```
ods rtf style=RSTYLERTF file="C:\Users\User\My Documents\APR\Ch2UnFmt.rtf";
```
        proc report data=sales nowd split="|" center missing
           style(column)=[just=c cellwidth=.8 in] out=prefmt;
        column REGION STORE BRANCH QTR4_2010 QTR1 QTR2;
           define REGION / order "Region"   style(column)=[cellwidth=.65 in];
           define STORE  / order "Store #"  style(column)=[cellwidth=.65 in];
           define BRANCH / order "Branch #" style(column)=[cellwidth=.7 in];
           define QTR4_2010 / "Quarter 4 2010" format=dollar10.;
           define QTR1 / "Quarter 1 2011" format=dollar10.;
           define QTR2 / "Quarter 2 2011" format=dollar10.;

        ** Sum Sales per store;
           break after STORE   / summarize suppress;
        run;

        ods rtf close;

        ods html;

        title;

        footnote;
```

For reference, Table 2.4 shows a partial print of the PROC REPORT output data set created in the PROC REPORT statement via "OUT=PREFMT". You can see that an extra column named _BREAK_ was produced. This automatic temporary variable would be created regardless of whether we had the BREAK statement; however, the _BREAK_ column would contain only blank values.

The "BREAK AFTER STORE /" statement by itself creates extra rows for which the value of _BREAK_ is "STORE" and the numeric columns, QTR4_2010, QTR1, QTR2 are summed.

The addition of the SUMMARIZE option allows the summary rows to be displayed in the printed report. The SUPPRESS option prevents the REGION and STORE values from being displayed on the summary lines of the printed report. You can see that the REGION and STORE values do still remain in the summary lines of the PROC REPORT data set.

Table 2.4 PROC REPORT Data Set "PREFMT"

REGION	STORE	BRANCH	QTR4_2010	QTR1	QTR2	_BREAK_
East	1001	1	43564	42555	47000	
East	1001	2	47235	47523	48102	
East	1001	3	50244	45999	46582	
East	1001	4	37665	37778	39624	
East	1001	.	178708	173855	181308	STORE
East	1002	1	50000	50248	49335	
East	1002	2	48045	46822	48000	
East	1002	3	60023	62410	62650	
East	1002	4	55000	56102	56109	
East	1002	5	52345	52650	54197	
East	1002	6	56444	57720	57800	
East	1002	7	59721	63000	65254	
East	1002	.	381578	388952	393345	STORE
East	1003	1	67443	66230	67503	
East	1003	2	69903	70085	73598	
East	1003	3	57900	60000	62900	
East	1003	4	58989	59222	58307	
East	1003	5	56112	57364	57647	
East	1003	.	310347	312901	319955	STORE
East	1007	1	51008	51008	52310	
East	1007	2	46733	43062	45564	
East	1007	3	61228	59838	62300	
East	1007	.	158969	153908	160174	STORE
East	1008	1	63328	63000	64274	
East	1008	2	61377	66461	70468	
East	1008	3	59821	59637	61272	
East	1008	.	184526	189098	196014	STORE

Implementation

Transforming Figure 2.1 Into Figure 2.2

The remainder of the chapter will add the code that transforms Figure 2.1 into Figure 2.2, one section at a time.

Displaying Region as a Line Above Each Report Page

As shown in Figure 2.3, the report displays Region as a line rather than as a report column.

Figure 2.3 - Display Region Before Report

⟶ **Region: East**

Store #	Branch #	Quarter 4 2010	Quarter 1 2011	Quarter 2 2011
1001	1	$43,564	$42,555	$47,000
	2	$47,235	$47,523	$48,102
	3	$50,244	$45,999	$46,582
	4	$37,665	$37,778	$39,624
		$178,708	*$173,855*	*$181,308*

Overview of Region Display

Displaying REGION as a line above each report page is accomplished with a BREAK statement and a COMPUTE block. A change in REGION is the designated BREAK before which a new page should be started. "Before" each page is the location for the COMPUTED Region LINE.

Key steps for displaying a Region line above each report page:

- Specify ORDER and NOPRINT options in the DEFINE statement for REGION.
- Insert a page break before each new value of REGION using a BREAK statement.
- Via a COMPUTE block, display the region specification before each page's report in the form "Region: "*REGION value*, and apply styles including left justification, bolded font, increased font size and blue foreground to the line.

54 *PROC REPORT by Example: Techniques for Building Professional Reports Using SAS*

Code to Make the Region Display in Figure 2.3

The following additions are made to Program 2.1 to display the region as a line.

```
** Specify ORDER and NOPRINT Options for REGION;   ❶

define REGION / order noprint;

** Insert Page Break Before each Region;

break before REGION / page;   ❷

** Display the Region Before each Page and Apply Styles;

compute before _page_;   ❸
   line "^S={foreground=blue font_weight=bold font_size=16 pt
         just=left} Region: " REGION $20.;   ❹
endcomp;
```

❶ While REGION is a report variable specified in the column statement, the goal is to print REGION in a line prior to the report, rather than as a column. Therefore, the NOPRINT option is applied in the DEFINE statement to prevent the column printing. The ORDER option is applied to allow for page BREAKs and COMPUTEd lines BEFORE REGION.

❷ To ensure that only one region's data appears on each page, a BREAK statement is used to start a new page for each new value of REGION.

❸ After ensuring each page contains only one region, the location for lines displaying region text is specified as "before _page_", which translates into "Display the region above the table on each page."

❹ Specifically, the region text should appear in the form "Region: "*Actual REGION value*.

To further highlight the region being reported, the font weight is set to bold, the font size is increased to 16 pt., and the font color is changed from the default black to blue. The line is also left justified.

Displaying Store and Branch Column Data in Bold Blue Font

Figure 2.4 shows the application of color and bold weight to STORE and BRANCH data. The style modifications are applied in these variables' DEFINE statements.

Figure 2.4 – Blue Font: Store and Branch Values

Region: East

Store #	Branch #	Quarter 4 2010	Quarter 1 2011	Quarter 2 2011
1001	1	$43,564	$42,555	$47,000
	2	$47,235	$47,523	$48,102
	3	$50,244	$45,999	$46,582
	4	$37,665	$37,778	$39,624
		$178,708	*$173,855*	*$181,308*

Code for Store and Branch Display

The highlighted code indicates the additions to the Program 2.1 STORE and BRANCH DEFINE statements.

```
** Modify Foreground and Font_weight for STORE and BRANCH Data;  ❶
define STORE  / order "Store #" style(column)=[cellwidth=.65 in
                               foreground=blue font_weight=bold];
define BRANCH / order "Branch #" style(column)=[cellwidth=.7 in
                               foreground=blue font_weight=bold];
```

❶ The columns are styled using the STYLE(COLUMN)= option, along with the CELLWIDTH=, FOREGROUND=, and FONT_WEIGHT= attribute specifications on the respective DEFINE statements. The CELLWIDTH= style attribute is used to define the STORE and BRANCH column sizes as .65 and .7 inches wide. The blue colored font ("foreground") and bold font weight are applied to STORE and BRANCH values with the FOREGROUND= and FONT_WEIGHT= specifications.

Note that the ORDER option is applied to both STORE and BRANCH (as was done for the unformatted report, Figure 2.1).

- Applying the ORDER usage option orders the stores and branches, prevents repeated values from being displayed and, importantly, prevents the numeric branch numbers from being assigned their default ANALYSIS. With an ANALYSIS usage, the branch numbers would be summed by the BREAK after the store / summarize statement.

How to Insert Arrows for Quick Reference to Sales Increases/Decreases

Increases/Decreases Arrows have been inserted so the reader can quickly identify if sales increased, decreased, or stayed the same from the previous quarter, as shown in Figure 2.5.

Figure 2.5 – Insert Arrows for Quick Reference

Region: East

Store #	Branch #	Quarter 4 2010	Quarter 1 2011		Quarter 2 2011	
1001	1	$43,564	$42,555	↓	$47,000	↑
	2	$47,235	$47,523	↑	$48,102	↑
	3	$50,244	$45,999	↓	$46,582	↑
	4	$37,665	$37,778	↑	$39,624	↑
		$178,708	*$173,855*	↓	*$181,308*	↑

Overview on Arrow Insertion

In this section, COMPUTE blocks are used to create new variables that represent the change in sales from quarter to quarter. The numeric differences are converted to arrows via a format. Color and font style is then applied to the arrows, including the summary line, via a color format used in CALL DEFINE statements. The key steps for inserting arrows to show changes from the previous quarter's sales include:

- Create the arrow and color formats to be applied to the sales differences.
- Add the COMPUTEd variable names DIR1 and DIR2 to the COLUMN Statement.
- Apply the Wingdings font_face and arrow format in the DIR1 and DIR2 DEFINE Statements.
- Using a COMPUTE Block:
    - Create the two difference variables DIR1 and DIR2.
        - DIR1 represents Quarter 1 2011 sales (QTR1) minus Quarter 4 2010 sales (QTR4_2010).
        - DIR2 represents Quarter 2 2011sales (QTR2) minus Quarter 1 2011 sales (QTR1).
    - Conditionally apply color to increases, decreases, and no change in CALL DEFINE statements so the summary line arrows receive the proper color application.

Code for Arrow Insertion

The following additions are made to Program 2.1 to create DIR1 and DIR2 and style these as symbols.

** Create Color and Arrow Formats; ❶

```
proc format;
  value arrow
    low - < 0 = "EA"x
    0 = "A2"x
    0 <- high = "E9"x;
  value color
    low - < 0 = red
    0 = black
    0 <- high = green;
run;
```

** Add Computed Variable Names DIR1 and DIR2 to the COLUMN Statement; ❷

```
proc report data=sales nowd split="|" center missing
                     style(column)=[just=c cellwidth=.8 in];
  columns REGION STORE BRANCH QTR4_2010 QTR1 DIR1 QTR2 DIR2;
```

** Add DIR1 and DIR2 DEFINE Statements. Apply Arrow Format and Wingdings Font; ❸

```
  define DIR1 / " " computed format=arrow.
                style(column)=[cellwidth=.6 in
                               font_face=wingdings];
  define DIR2 / " " computed format=arrow.
                style(column)=[cellwidth=.6 in
                               font_face=wingdings];
```

** Add Compute Blocks to Derive Difference Variables DIR1 And DIR2 and Style with Color. Format, including Summary Lines;

```
compute DIR1;
   DIR1=QTR1.sum-QTR4_2010.sum; ❹
```

58 PROC REPORT by Example: Techniques for Building Professional Reports Using SAS

```
        call define("DIR1","style","style={foreground=color.}");  ❺
    endcomp;
    compute DIR2;
        dir2=QTR2.sum-QTR1.sum;
        call define("DIR2","style","style={foreground=color.}");
    endcomp;
```

❶ Two formats are created. The arrow format contains ASCII hexadecimal (hex) codes ("EA","E9","A2") that will be rendered as arrows and circles once Wingdings font is applied in the DEFINE statements for the sales difference variables DIR1 and DIR2. Table 2.5 displays the Hex codes, and how they will be rendered in the RTF output after Wingdings font is applied.

Table 2.5 Arrow Format Specifying Hex Code and Corresponding Wingdings Representation

Hex Code	Wingdings Representation
A2	○
E9	↑
EA	↓

Visit http://dmcritchie.mvps.org/rexx/htm/fonts.htm for other example conversions of Hex Code to Wingdings.

The COLOR format will be used to apply either Red, Black, or Green to DIR1 and DIR2 in the CALL DEFINE statements.

❷ The computed variables are added to the COLUMN statement, since they must be report variables to show as columns.

❸ While the arrow format could have been applied in the conditional CALL DEFINE blocks, they are applied in the DEFINE statements, since all rows of the difference variables use the arrow format.

Based on the arrow format, negative sales differences are applied the "EA"x value (down arrow), 0s are applied the "A2"x value (open circle), and positive difference are applied the "E9"x value (up arrow). If the format is not applied, Wingdings symbols other than the desired ones will be displayed in the report.

Chapter 2: Formatting Highly Detailed Reports   59

❹ Per COMPUTE block rules, the analysis variables used to derive the computed variables are referenced by their compound names (*VARIABLE.statistic*) such as QTR1.sum and QTR4_2010.sum (sum is the default statistic for analysis variables).

❺ The CALL DEFINE statement is used to conditionally apply font color to the arrows and open circle using the COLOR format. The color format was applied within the COMPUTE block CALL DEFINE statement (versus in DIR1 and DIR2 DEFINE statements), so it would be applied to the summary line arrows as well (otherwise the summary line arrows display in blue colored font).

Increases and decreases use Wingdings font (to obtain the arrow symbol) and green versus red foregrounds, respectively. For no change (e.g. DIR1 or DIR2 = 0), a black open circle is shown.

How to Add Spanning Headers, Bottom Cell Borders, and Underlines

Figure 2.6 shows the addition of the "2011 Sales" header that spans over the 2011 quarterly sales columns and their corresponding arrows columns. Bottom cell borders and underlines are also added to help clarify sections of the report.

Figure 2.6 – Bottom Cell Borders and Text Underlines

Region: East

		2010 Sales	2011 Sales			
Store #	Branch #	Quarter 4	Quarter 1	Vs. Q4	Quarter 2	Vs. Q1
1001	1	$43,564	$42,555	↓	$47,000	↑
	2	$47,235	$47,523	↑	$48,102	↑
	3	$50,244	$45,999	↓	$46,582	↑
	4	$37,665	$37,778	↑	$39,624	↑
		$178,708	*$173,855*	↓	*$181,308*	↑

Highlights on Adding Spanning Headers, Borders, and Underlines

The Sales column headers are created in the COLUMN statement (versus DEFINE statements) to ensure

1. the ability to create spanning headers.
2. that these headers are displayed a level above the "Store #" and "Branch #" headers.

60 *PROC REPORT by Example: Techniques for Building Professional Reports Using SAS*

This section makes use of border control to add header bottom borders, and text decoration functionality to add header underlines. The example below shows the difference between a bottom border and a text underline.

Example of Bottom Border	Example of Text Underline
abcd	<u>abcd</u>

Bottom borders are added by inserting the inline formatting function borderbottomwidth as shown below.

"^{style [borderbottomwidth=1pt] *header text*}"

Text Underlines are added by inserting the inline formatting function textdecoration as shown below

"^S={textdecoration=underline} *header text*"

Specific steps in creating spanning headers include:

- In the COLUMN statement, add bottom cell borders and create a spanning header, "2011 Sales" to span over relevant columns.
- Add "Store #" and "Branch #" headers via the DEFINE statements so they are displayed a level below the sales headers. For these headers, text underlining is used rather than applying bottom borders to the table cells.
- Set the Sales variables' header text to null in the DEFINE statements since these headers are being labeled in the COLUMN statements.

Code for Adding Spanning Headers, Borders, and Underlines

The following additions are made to the Program 2.1 COLUMN and DEFINE statements to modify the column headers.

```
COLUMN region store branch
** Add Bottom Cell Borders And Create A Spanning Header;
    ("2010 Sales|^{style [borderbottomwidth=1 pt]Quarter 4}" QTR4_2010)
❶
    ("^{style [borderbottomwidth=1 pt] 2011 Sales}" ❷
        ("^{style [borderbottomwidth=1 pt] Quarter 1}" QTR1) ❸
```

```
             ("^{style [borderbottomwidth=1 pt] Vs. Q4}"     dir1)

             ("^{style [borderbottomwidth=1 pt] Quarter 2}" QTR2)

             ("^{style [borderbottomwidth=1 pt] Vs. Q1}"     dir2)
```

```
      );

** Underline Store # and Branch # Headers;  ❹

define store    / order "^S={textdecoration=underline}Store #"
                style(column)=[cellwidth=.65 in
                foreground=blue font_weight=bold];

define branch  / order "^S={textdecoration=underline}Branch #"
                style(column)=[cellwidth=.7 in
                foreground=blue font_weight=bold];
```

** Sales Column Headers are created in the COLUMN statement and set to null in DEFINE statements so variable names are not displayed; ❺

```
define QTR4_2010 /  ""  format=dollar10.;
define QTR1      /  ""  format=dollar10.;
define DIR1      / computed  ""  style(column)=[font_face=wingdings
                                   cellwidth=.6 in] format=arrow.;
define QTR2      /  ""  format=dollar10.;
define DIR2      / computed  ""  style(column)=[font_face=wingdings
                                   cellwidth=.6 in] format=arrow.;
```

❶ 2010 Sales Quarter 4 is a standalone item that is given a cell bottom border.

❷ A spanning header "2011 Sales" with a bottom border is created to encompass each 2011 Sales and each Increase/Decrease column, "Quarter 1," "Vs. Q4," "Quarter 2," and "Vs. Q1."

The "^{style [borderbottomwidth=1 pt] 2011 Sales}" code covers all columns within the outermost begin and end parentheses (i.e. encompasses all of the "❸" boxed code section).

❸ Within this spanning header, each individual 2011 Sales header is given a bottom border as well.

❹ The underlines for Store # and Branch # headers are applied by inserting the textdecoration inline formatting within the headers in the DEFINE statements. These headers could have been created in the COLUMN statement, but the preference for this report is to display these headers below the Sales header levels.

❺ Sales Column Headers are specified in the COLUMN statement and are therefore set to null (see gray highlighted code) in the DEFINE statements so the variable names are not displayed.

Adding Blank Columns to Make the Report More Legible

Figure 2.7 presents the report with added blank columns where white space is desired.

Figure 2.7 – Showing Added Blank Columns

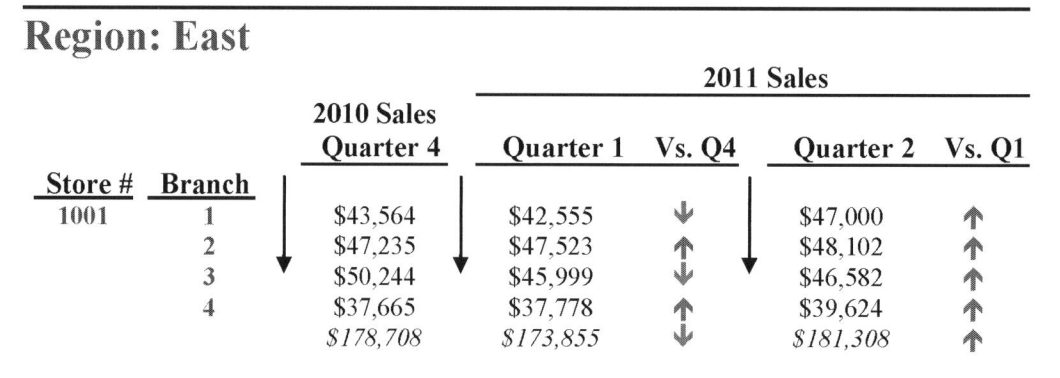

Region: East

		2010 Sales	2011 Sales			
		Quarter 4	Quarter 1	Vs. Q4	Quarter 2	Vs. Q1
Store #	Branch					
1001	1	$43,564	$42,555	↓	$47,000	↑
	2	$47,235	$47,523	↑	$48,102	↑
	3	$50,244	$45,999	↓	$46,582	↑
	4	$37,665	$37,778	↑	$39,624	↑
		$178,708	*$173,855*	↓	*$181,308*	↑

Overview of Adding Blank Columns

A DATA step is used to derive the new variable BLANK, which is set to null and specified as a PROC REPORT COLUMN variable where blank columns are desired.

- Create "BLANK" variable in data set.
- Add the new variable to the COLUMN statement in places for which a blank column is desired.
- Specify null label and the desired cell width for each blank column in the DEFINE statement.

Code for Adding Blank Columns

```
** Add "BLANK" variable to the SALES data set; ❶
data sales;
  set sales;
  blank = "";
run;
```

```
** Add BLANK to COLUMN Statement in Desired Locations;  ❷
COLUMN region store branch BLANK
** Add Bottom Cell Borders And Create A Spanning Header;
   ("2010 Sales|^{style [borderbottomwidth=1 pt]Quarter 4}" QTR4_2010)
   BLANK
      ("^{style [borderbottomwidth=1pt] 2011 Sales}"
         ("^{style [borderbottomwidth=1pt] Quarter 1}" QTR1)
         ("^{style [borderbottomwidth=1pt] Vs. Q4}"    dir1) BLANK
         ("^{style [borderbottomwidth=1pt] Quarter 2}" QTR2)
         ("^{style [borderbottomwidth=1pt] Vs. Q1}"    dir2)
   );
** Add BLANK DEFINE statement with null header and desired column width;  ❸
define BLANK / "" style(column)=[cellwidth=.15 in];
```

❶ The BLANK variable is added to the dataset to allow us to insert white space in the report where extra blank columns are desired.

❷ The COLUMN statement determines the column order in the report. Therefore, the BLANK variable is listed each time a blank column is desired within the COLUMN variable list. We use the BLANK variable three times in this report.

❸ If the labels are not set to null, the heading "BLANK" will appear as a column header in the report. If the cell width is not set, the default column width may produce a blank column wider/narrower than desired.

Note that the BLANK columns could have been created via COMPUTE blocks. For example, we could have used the following code to create three new report variables.

```
compute BLANK1 / char length=2;
   BLANK1 = " ";
endcomp;

compute BLANK2 / char length=2;
   BLANK2= " ";
endcomp;
```

64 *PROC REPORT by Example: Techniques for Building Professional Reports Using SAS*

```
compute BLANK3 / char length=2;
  BLANK3 = " ";
endcomp;
```

We would then reference the variable names BLANK1, BLANK2, and BLANK3 in the COLUMN statement, and use three DEFINE statements to apply null headers and desired cell widths.

Rather, for this example we took advantage of the automatic creation of aliases for the input data set variable BLANK that occurred because we referenced it more than once in the COLUMN statement. We can see that this automatic creation occurred by adding the LIST option to the PROC REPORT statement. LIST writes PROC REPORT code to the log, including some of the defaults we did not specify.

Using the LIST option in the PROC REPORT statement, the SAS log shows us that aliases _A1 and _A2 were created for BLANK.

```
PROC REPORT DATA=WORK.SALES LS=98   PS=55    SPLIT="|" CENTER MISSING ;
COLUMN   ( REGION STORE BRANCH
  blank
  ("2010 Sales|^{style [borderbottomwidth=1 pt]Quarter 4}" qtr4_2010 )
  blank=_A1
  ("^{style [borderbottomwidth=1pt] 2011 Sales}"
    ("^{style [borderbottomwidth=1pt] Quarter 1}"  qtr1 )
    ("^{style [borderbottomwidth=1pt] Vs. Q4}"dir1 )
  blank=_A2
    ("^{style [borderbottomwidth=1pt] Quarter 2}"  qtr2 )
    ("^{style [borderbottomwidth=1pt] Vs. Q1}"  dir2 )
) );
```

Also note, using this method for obtaining blank columns only required one DEFINE statement for the variable BLANK.

Style: Add a Blank Line After Each Summary Line

As a final step, summary lines are distinguished by adding color and white space as shown in Figure 2.8.

Figure 2.8 Style Summary Lines

Region: East

			2011 Sales			
		2010 Sales				
		Quarter 4	Quarter 1	Vs. Q4	Quarter 2	Vs. Q1
Store #	Branch #					
1001	1	$43,564	$42,555	↓	$47,000	↑
	2	$47,235	$47,523	↑	$48,102	↑
	3	$50,244	$45,999	↓	$46,582	↑
	4	$37,665	$37,778	↑	$39,624	↑
	→	*$178,708*	*$173,855*	↓	*$181,308*	↑
1002	1	$50,000	$50,248	↑	$49,335	↓
	2	$48,045	$46,822	↓	$48,000	↑
	3	$60,023	$62,410	↑	$62,650	↑
	4	$55,000	$56,102	↑	$56,109	↑
	5	$52,345	$52,650	↑	$54,197	↑
	6	$56,444	$57,720	↑	$57,800	↑
	7	$59,721	$63,000	↑	$65,254	↑
	→	*$381,578*	*$388,952*	↑	*$393,345*	↑

Highlights on Styling Summary Line and Adding a Blank Line

Key steps to highlighting sales totals include:

- Summarize with a BREAK statement, which was done for the unformatted report as well. In this example, we apply bold blue font to the summary lines.
- Further highlight the totals by adding a blank line after the last value of each Store #.

Code for Styling Summary Line and Adding a Blank Line

```
** Apply Bold Blue Font to Summary Lines;

break after STORE    / summarize suppress
style(summary)=[font_weight=bold foreground=blue]; ❶

** Add Blank Line after each store # section ❷;
```

```
compute after STORE;
  line " ";
endcomp;
```

❶ Specify that PROC REPORT summarize the analysis variables after STORE (i.e. each store number). BRANCH, a numeric variable would be an ANALYSIS variable by default, and therefore summarized, but the ORDER option overrides this.

We specify desired styles for the summary line with the style(summary)= option.

❷ Via a COMPUTE block, add a blank line after each STORE.

Chapter 2 Summary

The chapter demonstrated ways to turn a highly detailed report (Figure 2.1) into a manageable readable format (Figure 2.2) simply by applying various ODS and PROC REPORT features. Table 2.6 recaps the enhancements made to Figure 2.2 and the main programming modifications that were required.

Table 2.6 Chapter Summary

Enhancements	**Main Programming Modifications**
◆ Region was displayed as a line above each report page, rather than as a column on each page.	Specifying NOPRINT in the REGION DEFINE statement; BREAK Statement to create page break (before REGION / page), COMPUTE Block to create a Region line (before _page_).
◆ Store #s and Branch #s were displayed in bold blue font ("foreground").	Styled with Style(column) = options in DEFINE statements.
◆ Arrows were displayed after each 2011 Sales column to provide a quick reference to whether sales increased, decreased, or stayed the same from the previous quarter.	COMPUTED variables to derive quarterly sales differences; Formats were used to apply symbol type and color to the difference variables. Color was applied in CALL DEFINE statements to override the blue colored font of Summary Line symbols.
◆ A spanning header, "2011 Sales" was created and bottom borders and underlines were added to header cells.	COLUMN statement used to create spanning and nested headers with the use of parentheses. Inline formatting style functions were inserted to add borders and underlines.

Enhancements	Main Programming Modifications
◆ Summary lines, which show total Store sales, were brought out by applying blue font color and adding a blank line after each summary line, before the start of each new store number.	Summary lines were styled with the style(summary)= option; A COMPUTE block was used to add a blank line after each STORE.
◆ Blank columns were added after the following columns: Branch #, 2010 Sales Quarter 4, and Vs Q4.	A DATA step was used to create a new variable, BLANK. This was reported in desired locations by adding BLANK to the PROC REPORT COLUMN statement where blanks were desired. The desired cell width and a null header were applied in a DEFINE statement.

Figure 2.1 BEFORE Formatting

Region	Store #	Branch #	Quarter 4 2010	Quarter 1 2011	Quarter 2 2011
East	1001	1	$43,564	$42,555	$47,000
		2	$47,235	$47,523	$48,102
		3	$50,244	$45,999	$46,582
		4	$37,665	$37,778	$39,624
			$178,708	*$173,855*	*$181,308*
	1002	1	$50,000	$50,248	$49,335
		2	$48,045	$46,822	$48,000
		3	$60,023	$62,410	$62,650
		4	$55,000	$56,102	$56,109
		5	$52,345	$52,650	$54,197
		6	$56,444	$57,720	$57,800
		7	$59,721	$63,000	$65,254
			$381,578	*$388,952*	*$393,345*
	1003	1	$67,443	$66,230	$67,503
		2	$69,903	$70,085	$73,598
		3	$57,900	$60,000	$62,900
		4	$58,989	$59,222	$58,307
		5	$56,112	$57,364	$57,647
			$310,347	*$312,901*	*$319,955*
	1007	1	$51,008	$51,008	$52,310
		2	$46,733	$43,062	$45,564
		3	$61,228	$59,838	$62,300
			$158,969	*$153,908*	*$160,174*
	1008	1	$63,328	$63,000	$64,274
		2	$61,377	$66,461	$70,468
		3	$59,821	$59,637	$61,272
			$184,526	*$189,098*	*$196,014*
	1009	1	$60,044	$61,900	$64,528
			$60,044	*$61,900*	*$64,528*
	1011	1	$47,889	$42,500	$40,389
			$47,889	*$42,500*	*$40,389*
	1012	1	$39,698	$42,870	$42,999

Figure 2.2 AFTER Formatting

Region: East

Store #	Branch #	2010 Sales Quarter 4	2011 Sales Quarter 1	Vs. Q4	Quarter 2	Vs. Q1
1001	1	$43,564	$42,555	↓	$47,000	↑
	2	$47,235	$47,523	↑	$48,102	↑
	3	$50,244	$45,999	↓	$46,582	↑
	4	$37,665	$37,778	↑	$39,624	↑
		$178,708	*$173,855*	↓	*$181,308*	↑
1002	1	$50,000	$50,248	↑	$49,335	↓
	2	$48,045	$46,822	↓	$48,000	↑
	3	$60,023	$62,410	↑	$62,650	↑
	4	$55,000	$56,102	↑	$56,109	↑
	5	$52,345	$52,650	↑	$54,197	↑
	6	$56,444	$57,720	↑	$57,800	↑
	7	$59,721	$63,000	↑	$65,254	↑
		$381,578	*$388,952*	↑	*$393,345*	↑
1003	1	$67,443	$66,230	↓	$67,503	↑
	2	$69,903	$70,085	↑	$73,598	↑
	3	$57,900	$60,000	↑	$62,900	↑
	4	$58,989	$59,222	↑	$58,307	↓
	5	$56,112	$57,364	↑	$57,647	↑
		$310,347	*$312,901*	↑	*$319,955*	↑
1007	1	$51,008	$51,008	○	$52,310	↑
	2	$46,733	$43,062	↓	$45,564	↑
	3	$61,228	$59,838	↓	$62,300	↑
		$158,969	*$153,908*	↓	*$160,174*	↑
1008	1	$63,328	$63,000	↓	$64,274	↑

Chapter 3: Reporting Different Metrics Within a Column

Introduction 70
Example: Demographic and Baseline Characteristics Report 70
Goals for the Demographics and Baseline Characteristics Report 72
    Key Steps 72
Source Data 73
ODS Style Template Used 74
Programs Used 74
Implementation 74
Obtain Population Counts for Column Headers and Denominators 74
    Code for Obtaining Population Counts 75
Categorical Variables: Obtain Counts and Percentages 75
    Code for Obtaining Categorical Counts and Percentages 76
Continuous Variables: Descriptive Data 81
    Macro Code for Obtaining Descriptive Statistics 81
Create Final Table: Combine TABULATE and MEANS Results 85
    Code for Combing the Results 85
Produce the Report via PROC REPORT 89
    PROC REPORT Code 89

Chapter 3 Summary 91

Introduction

One reporting challenge arises when various data types need to be reported within a column. This is often the case in a Clinical Trials Demographics and Baseline Characteristics report, in which some of the raw data variables are categorical while others are continuous. The categories are reported as counts and percentages, while the continuous variables are summarized by a set of descriptive statistics such as the mean, median, standard deviation, and minimum and maximum values. In addition, the metrics require varying levels of precision and call for the summaries to line up visually by using indentation and/or decimal alignment.

Example: Demographic and Baseline Characteristics Report

For this example, patients are randomly assigned to one of two treatment groups, each group having a corresponding planned dose level of Study Medication (Dose Level 1 versus Dose Level 2). Four demographic and baseline characteristics are reported for each treatment group. Gender and Race are categorical variables, while Age and Weight are continuous variables. Figure 3.1 displays the report to be produced in this chapter.

Figure 3.1 Chapter 3 Report

Demographic and Baseline Characteristics

	Dose Group	
Parameter	Dose Level 1 (N=15)	Dose Level 2 (N=13)
Gender		
Female	5 (33.3%)	6 (46.2%)
Male	10 (66.7%)	7 (53.8%)
Race		
American Indian or Alaska Native	0	0
Asian	2 (13.3%)	1 (7.7%)
Black or African American	4 (26.7%)	5 (38.5%)
Hispanic or Latino	1 (6.7%)	1 (7.7%)
Native Hawaiian or Other Pacific Islander	0	0
White	6 (40.0%)	4 (30.8%)
Mixed Race	2 (13.3%)	1 (7.7%)
Missing	0	1 (7.7%)
Age		
n	15	13
Mean	50.7	54.9
Median	50.0	56.0
Standard Deviation	9.20	7.22
Min, Max	37, 69	42, 67
Weight (kg)		
n	15	13
Mean	76.889	75.671
Median	78.340	75.400
Standard Deviation	9.0920	7.9917
Min, Max	56.50, 89.00	65.20, 90.20

Goals for the Demographics and Baseline Characteristics Report

The goals for the Demographics and Baseline Characteristics report include:

- Reporting varying data types within a column, such as:
  - Categorical data (Gender and Race) versus Continuous data (Age and Weight).
    - Different metrics: counts and percentages for categorical data; n, mean, median, standard deviation, minimum and maximum for continuous variables.
- Application of desired precision with the help of macro variables.
  - Percentages are reported to one decimal.
  - Number of decimal places reported for continuous statistics depends on the specific variable and statistic being reported.
- Alignment of decimals and other alignment for the various metrics.
- For categorical variables, reporting all categories on the data collection form, even if there were no occurrences in the data.

Key Steps

The key steps taken to accomplish the goals include:

1. Obtaining the column N's (population counts) for display in the report column headers.
2. Use of the TABULATE and MEANS procedures to obtain the needed metrics and output data sets.
3. Modifying the TABULATE and MEANS output data sets to arrive at the common structure needed for the final output.
   - Applying rounding to numeric data
   - Converting numeric data to character strings
   - Deriving variables needed for a common data structure
   - Transposing the data to get treatment groups across the top.
4. Concatenating PROC TABULATE and PROC MEANS output once the data sets have a common structure.
5. Using ODS and PROC REPORT to style the report.

Source Data

There is one source data set named Ch3Demo. Ch3Demo contains a patient's demographic information, including the variables AGE, RACE, GENDER, and WEIGHTKG (weight in kilograms), each patient's identification number (SUBJID), and the treatment group to which they were assigned (TRT).

Tables 3.1 and 3.2 display the variable information and partial data for the data set "Ch3Demo".

Table 3.1 Ch3Demo Variable Information

#	Variable	Type	Len	Label
1	SUBJID	Char	8	Patient ID
2	TRT	Char	8	Treatment Group
3	RACE	Char	8	Race
4	GENDER	Char	8	Gender
5	WEIGHTKG	Num	8	Weight (kg)
6	AGE	Num	8	Age (years)

Table 3.2 Partial Ch3Demo Data

SUBJID	TRT	RACE	GENDER	WEIGHTKG	AGE
1	1	BA	M	80.00	44
3	1	WH	M	77.30	47
4	1	WH	M	82.40	50
5	1	WH	F	62.66	37
7	1	HL	M	68.60	48
8	1	MX	M	84.00	60
9	1	MX	M	88.60	55
10	1	WH	M	76.20	52

ODS Style Template Used

The report is produced in the Output Delivery System (ODS) Rich Text Format (RTF) destination. The Journal style template is used and specified prior to the PROC REPORT section in the statement

```
ods rtf style=journal file="{PATH\FILENAME}.rtf";
```

Programs Used

One program, Ch3Demo.sas, is used to create the Demographics and Baseline Characteristics Report.

Implementation

The remainder of the chapter describes the steps taken to create the Demographics and Baseline Characteristics report.

PROC TABULATE is used to obtain the counts and percentages for GENDER and RACE.

PROC MEANS is used to obtain descriptive statistics for the continuous variables AGE and WEIGHTKG, including the mean, median, standard deviation, minimum and maximum values and the number of observations on which the statistics are based.

Because the TABULATE and MEANs output require slightly different processing, these sections are demonstrated separately.

Obtain Population Counts for Column Headers and Denominators

The population counts are obtained for each treatment (dose level) group. The count for each population is saved into its own macro variable for later reporting of the population count in each treatment group column header.

Key steps for obtaining population counts include:

- Getting Population Counts via PROC FREQ.
- Storing Treatment Group Population Counts in Macro Variables &POP1 and &POP2.

Code for Obtaining Population Counts

```
** Get Data;
data demo;
   set sasuser.ch3demo;
run;

** Get Population Counts via PROC FREQ;
proc freq data= demo noprint;  ❶
  tables trt /out=pop;
run;

** Store Treatment Group Population Counts in Macro Variables;
data _null_;
  set pop;
  call symputx("POP" || TRT, COUNT);  ❷
run;

%put Population for trt1 = &pop1;
%put Population for trt2 = &pop2;
```

❶ A PROC FREQ is run to obtain the population counts for each treatment group.

❷ CALL SYMPUTX is used to pass the population counts for TRT=1 and TRT=2 into the macro variables, &POP1 and &POP2, respectively. The %PUT statements result in the log showing the following population counts:

```
Population for trt1 = 15
Population for trt2 = 13
```

Categorical Variables: Obtain Counts and Percentages

The key steps taken to prepare the categorical data for the final report include:

- Add PICTURE Format for Character Percent Alignment.
- Add formats to be used in PROC TABULATE.

- Use the TABULATE procedure to obtain the categorical variable counts and percentages for all categories specified in preloaded formats. Output the results to data sets.
- Create a character string that concatenates count and percent in the form n (xx.x). The percentages are rounded prior to being converted to character type.
- Derive variable VARNAME, needed for the final output structure.
- Transpose the TABULATE output data sets to get treatment groups across the top. SUBCAT, another variable needed for the final output structure, is obtained in this step.

Code for Obtaining Categorical Counts and Percentages

```
**   Create needed formats; ❶
proc format;

  **  Add PICTURE Format for Character Percentage Alignment;
  picture pctdec (round)
    0 - 1000 = "0009.9%)" (prefix="(")
    Other    = " ";

  **  Add Formats to be preloaded in the TABULATE procedure;
  **  We will use PRELOADFMT to produce rows with zero observations;
  value $sex
    "F"  = "Female"
    "M"  = "Male";


  value $race
      "AI" = "American Indian or Alaska Native"
      "AS" = "Asian"
      "BA" = "Black or African American"
      "WH" = "White"
      "HL" = "Hispanic or Latino"
      "MX" = "Mixed Race"
```

```
      "NH" = "Native Hawaiian or Other Pacific Islander"
      " " = "Missing";
run;

%MACRO TAB(tabvar=, indat=demo, fmt=, debug=N);
```
/** A Macro for Character Variables **/
/** Obtain Counts and Percentages **/ ❷
```
   proc tabulate data=&indat missing
                 out=m_&tabvar.1(drop=_type_ _page_ _table_);
     by trt;
     class &tabvar  / preloadfmt;
     table &tabvar, n*f=8. pctn / misstext="0" printmiss;
     format &tabvar &fmt..;
   run;

   data m_&tabvar.2;
     length &tabvar varname $16 pctc $30;
     set m_&tabvar.1;
```
/** Create Character String for Count and Percentage **/ ❸
```
     if pctn_0=0 then pctc="0";
     else pctc = strip(put(n,3.)|| " " ||put(pctn_0,pctdec.));
```
/** Derive Variable Needed for Final Output Structure **/
```
     varname= upcase("&tabvar"); ❹
   run;
```
/** Transpose the TABULATE Data **/
```
   proc sort data=m_&tabvar.2 out=m_&tabvar.3;
     by varname &tabvar trt;
   run;
```

```
   proc transpose data=m_&tabvar.3
       out=m_&tabvar.4(drop=_NAME_ rename=(&tabvar=subcat))
       prefix=trt;   ❺
   by varname &tabvar;
   id trt;
   var pctc;
run;

/** Delete Unnecessary Data Sets After Debugging is Complete **/
   %if &debug=Y %then
      %do;
         proc sql;
            drop table m_&tabvar.1,
                       m_&tabvar.2,
                       m_&tabvar.3;
         quit;
      %end;
%MEND TAB;
%TAB(tabvar=gender, indat=demo, fmt=$sex,  debug=Y)
%TAB(tabvar=race  , indat=demo, fmt=$race, debug=Y)
```

❶ A picture format is created for standardizing the character percent format. The (ROUND) option is added to the picture statement so percentages will be rounded before a format is applied. The picture format specification

```
     0 - 1000 ="0009.9%)" (prefix="(")
```

easily decimal aligns the percentages, adds a % sign, and surrounds the string in parentheses without leading and trailing blanks. The example below shows that the Race percentages are decimal aligned.

Race

American Indian or Alaska Native	0		0	
Asian	2	(13.3%)	1	(7.7%)
Black or African American	4	(26.7%)	5	(38.5%)
Hispanic or Latino	1	(6.7%)	1	(7.7%)

The $SEX and $RACE formats are created to pre-specify (preload) the desired categories to be shown in the TABULATE procedure output.

For each categorical variable (SEX and RACE in this example), the following steps are implemented via a macro call.

❷ PROC TABULATE is run and the counts and percentages are saved to an output dataset containing the name of the variable.

The guideline for this report is to display all possible categories shown on the data collection form even when there are no occurrences in the actual data. The following specifications in PROC TABULATE allow us to obtain the counts and percentages for all possible categories.

- The MISSING option in the PROC TABULATE statement includes an observation that contains a missing value for a class variable. Without the MISSING option, observations with missing values for class variables would not be included in the analysis.
- The PRELOADFMT in the CLASS statement, along with PRINTMISS in the TABLE statement, are used to display all possible combinations of formatted class variable values. This allows us to show when there are zero cases of a formatted value, for example, in our data set, the race="American Indian or Alaskan Native" category does not occur.
- The format statement specifies the format to apply to each class variable.
- MISSTEXT="0" specifies that the text "0" should be printed rather than the default period (".") to represent missing values.

❸ The character string PCTC is created. PCTC contains count and percentage in the form *n (xx.x)*. The character percentage portion *(xx.x)* is created with the picture format pctdec. The string will be decimal aligned in the RTF output when the column style just=d is used.

❹ Because the individual TABULATE output data sets will be concatenated later on, the identification variable VARNAME is added to each output data set. Table 3.3 shows the PROC PRINT for RACE at this point in the process.

Table 3.3 TABULATE Data for RACE – Pre-Transpose

varname	TRT	race	N	PctN_0	pctc
RACE	1	Missing	.	0.0000	0
RACE	1	American Indian or Alaska Native	.	0.0000	0
RACE	1	Asian	2	13.3333	2 (13.3%)
RACE	1	Black or African American	4	26.6667	4 (26.7%)
RACE	1	Hispanic or Latino	1	6.6667	1 (6.7%)
RACE	1	Mixed Race	2	13.3333	2 (13.3%)
RACE	1	Native Hawaiian or Other Pacific Islander	.	0.0000	0
RACE	1	White	6	40.0000	6 (40.0%)
RACE	2	Missing	1	7.6923	1 (7.7%)
RACE	2	American Indian or Alaska Native	.	0.0000	0
RACE	2	Asian	1	7.6923	1 (7.7%)
RACE	2	Black or African American	5	38.4615	5 (38.5%)
RACE	2	Hispanic or Latino	1	7.6923	1 (7.7%)
RACE	2	Mixed Race	1	7.6923	1 (7.7%)
RACE	2	Native Hawaiian or Other Pacific Islander	.	0.0000	0
RACE	2	White	4	30.7692	4 (30.8%)

❺ The data are transposed so that the categorical variable is reported with Treatment Groups 1 and 2 across the top. The &tabvar variable (RACE in this case) is renamed to "SUBCAT", one of the needed variables for the final output data structure. Table 3.4 shows the PROC PRINT for RACE, after the transpose.

Table 3.4 Transposed Data for RACE

varname	subcat	trt1	trt2
RACE	Missing	0	1 (7.7%)
RACE	American Indian or Alaska Native	0	0
RACE	Asian	2 (13.3%)	1 (7.7%)
RACE	Black or African American	4 (26.7%)	5 (38.5%)
RACE	Hispanic or Latino	1 (6.7%)	1 (7.7%)
RACE	Mixed Race	2 (13.3%)	1 (7.7%)

varname	subcat	trt1	trt2
RACE	Native Hawaiian or Other Pacific Islander	0	0
RACE	White	6 (40.0%)	4 (30.8%)

The other category variables are run through the TAB macro as well to generate output that looks similar in structure to the RACE output.

Continuous Variables: Descriptive Data

The MEANS macro is used to generate the needed statistics via PROC MEANS and to perform additional processing that structures the MEANS data similarly to the TABULATE data.

The key steps taken to prepare the MEANS data for the final report include:

- Use of the MEANS procedure to obtain the continuous variable descriptives and output data sets.
- Round output data to the appropriate decimal places.
- Convert numeric metrics to character versions.
- Concatenate minimum and maximum values in the form min, max.
- Derive variable needed for the final output structure, VARNAME.
- Transpose the MEANS output data sets to get treatment groups across the top. SUBCAT, another variable needed for the final output structure, is obtained in this step.

Macro Code for Obtaining Descriptive Statistics

```
%MACRO MEANS(meanvar=,indata=demo,rawdec=,rnddec=,debug=N);  ❶

   /** Use PROC MEANS to Obtain Continuous Variable Descriptives **/

   proc means data=&indata noprint;
      class trt;
      var &meanvar;
      output out=m_&meanvar.1(where=(_TYPE_ ne 0))
             n=N p50=MED mean=MEAN stddev=SD min=MIN max=MAX;  ❷
   run;
```

```
data m_&meanvar.2;
   length varname $16 nc meanc medc sdc minmaxc $50;
   set m_&meanvar.1(drop=_TYPE_ _FREQ_);
/** Round Mean, Median, and SD **/ ❸
   if mean ne . then mean = round(mean,.&rnddec);
   if med  ne . then med  = round(med ,.&rnddec);
   if sd   ne . then sd   = round(sd  ,.0&rnddec);
/** Create Character Versions of Statistics **/ ❹
   meanc   = strip(put(mean,12.%eval(&rawdec+1)));
   medc    = strip(put(med ,12.%eval(&rawdec+1)));
   sdc     = strip(put(sd  ,12.%eval(&rawdec+2)));
   nc      = strip(put(n   ,12.));
   minc    = strip(put(min ,12.&rawdec));
   maxc    = strip(put(max ,12.&rawdec));
/** Concatenate minimum and maximum values in the form min, max **/
   minmaxc = strip(minc) || ", " || strip(maxc); ❺
/** Derive Variable Needed for Final Output Structure **/
   varname = upcase("&meanvar"); ❻
run;

/** Transpose the MEANS Data **/
proc sort data=m_&meanvar.2 out=m_&meanvar.3;
   by varname trt;
run;

proc transpose data=m_&meanvar.3
   out=m_&meanvar.4
       prefix=trt name=subcat; ❼
```

```
     by varname;
     id trt;
     var nc meanc medc sdc minmaxc;
   run;

   /** Delete Unnecessary Data Sets After Debugging is Complete **/
   %if &debug=Y %then
     %do;
       proc sql;
         drop table m_&meanvar.1,
                    m_&meanvar.2,
                    m_&meanvar.3;
       quit;
     %end;
%MEND MEANS;

%MEANS(meanvar=AGE,     indata=demo,rawdec=0,RNDDEC=1,  debug=Y)
%MEANS(meanvar=WEIGHTKG,indata=demo,rawdec=2,RNDDEC=001,debug=Y)
```

❶ Macro parameters include:

Macro Variable	Description
&MEANVAR	the variable for which statistics are being obtained (e.g., AGE, WEIGHTKG)
&INDATA	the input dataset name
&RAWDEC	represents the maximum number of decimal places reported in the raw data for that variable
&RNDDEC	contains the needed precision for the round functions
&DEBUG	Specify if debugging is complete. If so, interim unneeded data sets will be deleted

❷ The needed PROC MEANS statistics are saved to an output dataset.

84   PROC REPORT by Example: Techniques for Building Professional Reports Using SAS

❸ Means, medians, and standard deviations are rounded using the ROUND function along with &RNDDEC to specify the rounding unit.

For this report, each variable's means and medians are reported to one decimal place beyond the maximum amount of decimal places found in the raw data for that variable. Standard deviations are reported to two decimal places beyond the maximum amount of decimal places found in the raw data for that variable.

For AGE, the maximum number of decimal points in the raw data is 0, therefore &RAWDEC is set to 0 in the macro call. Setting &RNDEC to 1 in the macro call arrives at the correct rounding since MEAN and MED are rounded to ".&RNDDEC" and SD is rounded to ".0&RNDDEC".

The rounding unit for AGE becomes the shaded portion of the following code upon macro resolution:

```
MPRINT(MEANS):   if mean ne . then mean = round(mean, .1);
MPRINT(MEANS):   if med  ne . then med  = round(med , .1);
MPRINT(MEANS):   if sd   ne . then sd   = round(sd  , .01);
```

❹ Character versions of each statistic are created using PUT functions and the macro variable &RAWDEC to apply the desired format.

For AGE, the formats using &RAWDEC resolve to the following:

```
MPRINT(MEANS):   meanc = strip(put(mean, 12.1));
MPRINT(MEANS):   medc  = strip(put(med , 12.1));
MPRINT(MEANS):   sdc   = strip(put(sd  , 12.2));
MPRINT(MEANS):   nc    = strip(put(n   , 12.)) ;
MPRINT(MEANS):   minc  = strip(put(min , 12.0));
MPRINT(MEANS):   maxc  = strip(put(max , 12.0));
```

❺ Minimum and maximum values are concatenated and reported in the form *min, max*.

❻ The variable VARNAME is added to the output data set so that variables and their corresponding results can be identified once the individual TABULATE and MEANS data sets are combined. Table 3.5 shows the PROC PRINT for AGE at this point in the process.

Chapter 3: Reporting Different Metrics Within a Column 85

Table 3.5 Print of AGE Data – Pre-Transpose

varname	TRT	nc	meanc	medc	sdc	minmaxc
AGE	1	15	50.7	50.0	9.20	37, 69
AGE	2	13	54.9	56.0	7.22	42, 67

❼ The MEANS output data set is transposed so that treatment groups are reported across the top. The name= option renames the default transpose variable "_NAME_" to "SUBCAT", one of the needed variables for the final output data structure. Table 3.6 shows the transposed data for AGE.

Table 3.6 Print of Transposed AGE Data

varname	subcat	trt1	trt2
AGE	nc	15	13
AGE	meanc	50.7	54.9
AGE	medc	50.0	56.0
AGE	sdc	9.20	7.22
AGE	minmaxc	37, 69	42, 67

Create Final Table: Combine TABULATE and MEANS Results

Once the TABULATE and MEANS data are in the same structure, the data are appended. Formats needed to create new variables are added prior to the DATA step.

Code for Combing the Results

```
** Add Formats needed for creation of report variables; ❶

{In the PROC FORMAT section…}

** Used with NOPRINT, for Ordering Variables;
invalue varord
   "GENDER"   = 1
   "RACE"     = 2
```

```
    "AGE"      = 3
    "WEIGHTKG" = 4;
```

** Used with NOPRINT, for Ordering Variables;

```
invalue subctord
    "AI"      = 1
    "AS"      = 2
    "BA"      = 3
    "HL"      = 4
    "NH"      = 5
    "WH"      = 6
    "MX"      = 7
    "F"       = 8
    "M"       = 9
    "nc"      = 10
    "meanc"   = 11
    "medc"    = 12
    "sdc"     = 13
    "minmaxc" = 14
    " "       = 15;
```

** Character to Character Format;

** This will format both levels of character variables and statistics;

```
value $subcat
    "AI"   = "American Indian or Alaska Native"
    "AS"   = "Asian"
    "BA"   = "Black or African American"
    "WH"   = "White"
    "HL"   = "Hispanic or Latino"
```

```
    "MX"      = "Mixed Race"

    "NH"      = "Native Hawaiian or Other Pacific Islander"

    " "       = "Missing"

    "F"       = "Female"

    "M"       = "Male"

    "nc"      = "n"

    "meanc"   = "Mean"

    "medc"    = "Median"

    "sdc"     = "Standard Deviation"

    "minmaxc" = "Min, Max";

** Concatenate the TABULATE and MEANS Data Sets; ❷
data all ;
  length trt1 trt2 newcat $100;
  set m_gender4   (in=ingend)
      m_race4     (in=inrace)
      m_age4      (in=inage)
      m_weightkg4 (in=inwt);

  ** Create New Reporting Variables; ❸
  ** Order Variables;
  varord = input(varname,varord.);
  subctord = input(subcat,subctord.);

  ** Display Variable;
  newcat = strip(put(subcat,$subcat.));
run;
```

❶ Formats needed prior to running the DATA step are added to the PROC FORMAT section.
❷ The MEANS and TABULATE data sets are combined.

❸ Three new variables are derived. VARORD and SUBCTORD are created to assign numeric order to character values. These will be used to order the rows in the report. NEWCAT contains row labels to be presented in the report.

Table 3.7 shows the concatenated TABULATE and MEANS data.

Table 3.7 Combined TABULATE and MEANS Data

varord	subctord	varname	newcat	trt1	trt2
1	8	GENDER	Female	5 (33.3%)	6 (46.2%)
1	9	GENDER	Male	10 (66.7%)	7 (53.8%)
2	15	RACE	Missing	0	1 (7.7%)
2	1	RACE	American Indian or Alaska Native	0	0
2	2	RACE	Asian	2 (13.3%)	1 (7.7%)
2	3	RACE	Black or African American	4 (26.7%)	5 (38.5%)
2	4	RACE	Hispanic or Latino	1 (6.7%)	1 (7.7%)
2	7	RACE	Mixed Race	2 (13.3%)	1 (7.7%)
2	5	RACE	Native Hawaiian or Other Pacific Islander	0	0
2	6	RACE	White	6 (40.0%)	4 (30.8%)
3	10	AGE	n	15	13
3	11	AGE	Mean	50.7	54.9
3	12	AGE	Median	50.0	56.0
3	13	AGE	Standard Deviation	9.20	7.22
3	14	AGE	Min, Max	37, 69	42, 67
4	10	WEIGHTKG	n	15	13
4	11	WEIGHTKG	Mean	76.889	75.671
4	12	WEIGHTKG	Median	78.340	75.400
4	13	WEIGHTKG	Standard Deviation	9.0920	7.9917
4	14	WEIGHTKG	Min, Max	56.50, 89.00	65.20, 90.20

The VARNAME values you see in Table 3.7 will be modified to display what is shown in Figure 3.1 by applying a format during the PROC REPORT phase.

Produce the Report via PROC REPORT

Key Steps for producing the report via PROC REPORT include:

- Adding the format to be used in PROC REPORT.
- Specifying RTF style and RTF output file name.
- Overriding Journal style's default by applying header, line, and column styles in the PROC REPORT statement.
- Creating and underlining the spanning header.
- Creating row headers via COMPUTE block LINE statements.

PROC REPORT Code

```
**   Add Formats Needed for Row Headers;   ❶
{In the PROC FORMAT section…}
  value $var
    "GENDER"   = "Gender"
    "RACE"     = "Race"
    "AGE"      = "Age"
    "WEIGHTKG" = "Weight (kg)";

**   Specify RTF Style and Output File Names;
ods _all_ close;
ods escapechar = "^";
options nodate nonumber orientation=portrait;
ods rtf style=journal file="C:\Users\User\My Documents\APR\Ch3.rtf";   ❷
title j=center h=10 pt "Demographic and Baseline Characteristics";

**   Apply Header, Line, and Column Styles in PROC REPORT Statement;
proc report data=all nowd center split="|"
     style(header)=[font_weight=bold indent=.65 in asis=on]
     style(lines) =[just=l font_weight=bold font_face=arial
```

```
                    font_size=9.5 pt]
       style(column)=[just=d cellwidth=1.6 in];  ❸

** Create and Underline Spanning Header;
   columns varord varname subctord newcat
           ("^S={textdecoration=underline}Dose Group" TRT1 TRT2);  ❹

** Use ORDER ORDER=INTERNAL NOPRINT to Order Rows and Suppress
Printing;
** For Ordering, the variables must be left-most in the COLUMNS
statement;
   define varord   / order order=internal noprint;
   define varname  / order noprint;
   define subctord / order order=internal noprint;
   define newcat   / "Parameter" order order=internal
                     style(header)=[just=l indent=0 in]
                     style(column)=[just=l cellwidth=2.8 in
                     indent=.25 in];
   define TRT1     / "Dose Level 1|         (N=&POP1)";
   define TRT2     / "Dose Level 2|         (N=&POP2)";

** Create Row Headers via COMPUTE Block LINE Statements;
   compute before varname;  ❺
     line " ";
     line varname $var16.;
   endcomp;
run;
ods _all_ close;
title;
footnote;
ods html;
```

❶ The format **$VAR**, which will be used to create row headers, is added to the PROC FORMAT section.

❷ The RTF destination is opened and the ODS style template Journal is specified as the report style.

❸ Report styles are specified in the PROC REPORT statement to override some of the default Journal style specifications.

- The **style(header)**= portion requests that column headers are bolded and indented. Since INDENT specifies indentation for only the first line of output, ASIS=ON is used to preserve leading spaces added to a second level header following a split character (e.g., Dose Level 1| (N=&POP1)").
- The **style(lines)**= option sets the style element for LINE statements in compute blocks. We want our row headers (Gender, Race, Age, Weight (kg)) to be left justified and to display bolded, Arial, 9.5 pt. font.
- The **style(column)**= portion of code requests that the column data be decimal aligned and have a column width of 1.6 inches.

Note that in the DEFINE statement for the NEWCAT column, the header and column styles are overridden. For this column, we want the "Parameter" header and data to be left justified and the column to have a greater width.

❹ Report columns are specified in the order they should be used for the report (some are used for ordering rows, some are displayed). The code ("^S={textdecoration=underline}Dose Group" TRT1 TRT2) creates an underlined spanning header "Dose Group" over the two treatment group columns.

❺ The row headers, Gender, Race, Age, and Weight (kg) are created via a COMPUTE block LINE statement which displays the value of VARNAME with the $var16 format applied.

Chapter 3 Summary

This chapter showed an example of how to develop a report that starts with various data types and reports different metrics within a column.

- PROC TABULATE and PROC MEANS were used to obtain the statistics for this example. However, the reader should be aware that output from many SAS procedures can be modified in the same manner to obtain the desired result.
- Output from the procedures was saved to SAS data sets and data processing was performed to the variables prior to feeding the data into the report.
- Numeric processing (i.e. rounding of statistics) was performed first.

- Next, the numeric values were converted to character strings for display in the desired format.
- Modifications of the output data set structures were also employed.
- Data was transposed and variables across data sets were given common names.
- Macros and macro variables were used throughout the process to make the restructuring of data more dynamic.
- Once a common structure across datasets was obtained, output data sets were combined.
- PROC REPORT was used as a means to easily transform the final data into a professionally styled report.
- A spanning header was added and column widths were adjusted.
- The ORDER option allowed for the insertion of blank rows between variables and row headers via a COMPUTE block.
- ODS capabilities were incorporated to indent and underline headers and further improve the look of the report.

Figure 3.1 Chapter 3 Report

Demographic and Baseline Characteristics

| | Dose Group ||
Parameter	Dose Level 1 (N=15)	Dose Level 2 (N=13)
Gender		
Female	5 (33.3%)	6 (46.2%)
Male	10 (66.7%)	7 (53.8%)
Race		
American Indian or Alaska Native	0	0
Asian	2 (13.3%)	1 (7.7%)
Black or African American	4 (26.7%)	5 (38.5%)
Hispanic or Latino	1 (6.7%)	1 (7.7%)
Native Hawaiian or Other Pacific Islander	0	0
White	6 (40.0%)	4 (30.8%)
Mixed Race	2 (13.3%)	1 (7.7%)
Missing	0	1 (7.7%)
Age		
n	15	13
Mean	50.7	54.9
Median	50.0	56.0
Standard Deviation	9.20	7.22
Min, Max	37, 69	42, 67
Weight (kg)		
n	15	13
Mean	76.889	75.671
Median	78.340	75.400
Standard Deviation	9.0920	7.9917
Min, Max	56.50, 89.00	65.20, 90.20

Chapter 4: Lesion Data Quality Report— COMPUTE Blocks

Introduction 96

Example: Lesion Data Quality Report 96

Goals for Creating the Lesion Data Quality Report 98

    Key Steps 99

Source Data 99

ODS Style Template Used 100

Programs Used 101

Implementation 101

    COMPUTE Block Variables: DATA Step (Temporary) Versus REPORT (COLUMN Statement) Variables 101

ORDER by and Print Subject ID on Every Row with Greying Font 102

    Program for Subject ID Display 102

Identify Potential Data Issues 107

    Code for Displaying Potential Data Issues 108

Final Formatting: Create Spanning Headers 119

Chapter 4 Summary 120

Introduction

Our task is to Quality Control (QC) newly arrived data. As programmers, we often need to identify and handle potential data issues prior to producing a client report. Just a few examples of potential data issues include duplicated information, inconsistencies in reported data, and missing and invalid data values. A preliminary data quality report can help reveal cases which require some type of action—for example, data which needs to be checked back to the original source, or simply cases for which data handling rules need to be verified. PROC REPORT is a flexible tool to identify, report, and resolve potential data issues before they end up in the final report.

Example: Lesion Data Quality Report

In this simplified clinical programming example, quality reports are produced to check patients' lesion data. At each patient's Screening visit, which takes place prior to the patient's receipt of study medication, up to four lesions are scanned and lesion sizes are assessed. The same lesions that were scanned at the Screening visit are scanned at every post-Screening visit in order to determine if each lesion is responding to the study medication—in other words, to determine if the lesion is shrinking. Visits following the Screening visit are designated by Cycle number, where the Cycle 1 represents the first post-Screening visit, Cycle 2 represents the second post-Screening visit, and so on.

A report is produced to check the lesions being scanned at each visit, so appropriate comparisons across visits can be made. Figure 4.1 highlights the following potential data issues:

- Cases in which there are multiple scans and/or conflicting results for the same lesion at a particular visit.
- Visits which have a different number of lesions from the Screening visit

Figure 4.2 is the same report, demonstrating a case for which a subject has one or more post-Screening visits, yet no Screening visit (see Subject 1003). For this type of situation, a superscript "a" is attached to the Subject ID. The superscript corresponds to the footnote, "Subject has no Screening Visit."

Figure 4.1 Potential Data Issues Report

Potential Data Issues Report

| Lesion Detail ||||| POTENTIAL DATA ISSUES ||||
|||||| Within Visit || Lesion Count Differs from Screening ||
Subject ID	Visit	Lesion ID	Assessment Date	Lesion Size (mm)	Multiple Dates Per Lesion ID	Multiple Sizes Per Lesion ID	More Lesions	Fewer Lesions
1001	Screening	L1	10OCT2010	10				
1001		L2	10OCT2010	12				
1001		L3	10OCT2010	12				
1001		L4	10OCT2010	11				
1001		*Lesion Count = 4*						
1001	Cycle 1	L1	19DEC2010	9				
1001		L2	19DEC2010	10				
1001		L3	19DEC2010	10				
1001				11		✓		
1001		L4	19DEC2010	10				
1001			20DEC2010	10	✓			
1001		*Lesion Count = 6*					✓	
1001	Cycle 2	L1	11FEB2010	8				
1001		L4	11FEB2010	9				
1001		*Lesion Count = 2*						✓
1001	Cycle 3	L1	21APR2010	8				
1001		L2	21APR2010	7				
1001		L3	21APR2010	7				
1001		L4	21APR2010	7				
1001		*Lesion Count = 4*						
1001	Cycle 4	L1	01JUL2010	8				

[a] Subject has no Screening Visit

Figure 4.2 Potential Data Issues Report with Superscript Example (See Subject 1003)

Potential Data Issues Report

Subject ID	Visit	Lesion ID	Assessment Date	Lesion Size (mm)	Multiple Dates Per Lesion ID	Multiple Sizes Per Lesion ID	More Lesions	Fewer Lesions
		Lesion Detail			**Within Visit**		**Lesion Count Differs from Screening**	
1001		L4	16DEC2010	5				
1001		Lesion Count = 4						
1001	Cycle 9	L1	28JAN2011	21				
1001		L2	28JAN2011	7				
1001		L3	28JAN2011	5				
1001		L4	28JAN2011	5				
1001		Lesion Count = 4						
1001	Cycle 10	L1	09MAR2011	4				
1001		L2	09MAR2011	8				
1001		L3	09MAR2011	3				
1001		L4	09MAR2011	9				
1001		Lesion Count = 4						
1002	Screening	L1	01JAN2011	6				
1002		Lesion Count = 1						
1003[a]	Cycle 2	L1	01MAR2011	4				
1003[a]		Lesion Count = 1						✓

[a] Subject has no Screening Visit

Goals for Creating the Lesion Data Quality Report

The goals for producing the Lesion Data Quality report are:

- Use of PROC REPORT to identify and report potential data issues present in lesion data, including:
    - Multiple and/or conflicting scan dates/lesion measurements within a visit.
    - A different number of lesions than were scanned at Screening visit.

Key Steps

The key steps taken to produce the data quality reports include:

- Application of ORDER and COMPUTED variable usage options to achieve the reports.
- Developing COMPUTE block DATA step variables to retain needed information and REPORT variables to report the information.
- Use of PROC REPORT output data sets to clarify COMPUTE block operations.
- Use of CALL DEFINE statements to highlight potential data issues.

Source Data

There is one source data set, Ch4Lesn, which contains variables representing:

- Subject ID
- Character Visit Description (Screening, Cycle 1, Cycle 2, Cycle 3, etc.),
- Numeric Visit Code (0, 1, 2 3, etc.)
- Character Lesion ID (L1, L2, L3, L4)
- Numeric Lesion ID (1, 2, 3, 4)
- Lesion Size in Millimeters, and
- Lesion Scan Date.

Tables 4.1 and 4.2 display the variable information and a partial print of the data set Ch4Lesn.

Table 4.1 Ch4Lesn Variable Information

#	Variable	Type	Len	Format	Informat	Label
1	SUBJECT	Char	8			Subject ID
2	VISIT	Char	16	$16.	$16.	Visit (Char)
3	VisitN	Num	8			Visit (Num)
4	LesC	Char	30	$30.	$30.	Lesion ID (Char)
5	LesN	Num	8			Lesion ID (Num)
6	SizeMM	Num	8			Lesion Size (mm)
7	ScanDT	Num	8	DATE9.		Scan Date

Table 4.2 Partial Ch4Lesn Data

SUBJECT	VISIT	VisitN	LesC	LesN	SizeMM	ScanDT
1001	Screening	0	L1	1	10	10OCT2010
1002	Screening	0	L1	1	6	01JAN2011
1003	Cycle 2	2	L1	1	4	01MAR2011
1001	Screening	0	L2	2	12	10OCT2010
1001	Screening	0	L3	3	12	10OCT2010
1001	Screening	0	L4	4	11	10OCT2010
1001	Cycle 1	1	L1	1	9	19DEC2010
1001	Cycle 1	1	L2	2	10	19DEC2010
1001	Cycle 1	1	L3	3	11	19DEC2010
1001	Cycle 1	1	L4	4	10	19DEC2010
1001	Cycle 1	1	L3	3	10	19DEC2010
1001	Cycle 1	1	L4	4	10	20DEC2010
1001	Cycle 2	2	L1	1	8	11FEB2010
1001	Cycle 2	2	L4	4	9	11FEB2010
1001	Cycle 3	3	L1	1	8	21APR2010
1001	Cycle 3	3	L2	2	7	21APR2010
1001	Cycle 3	3	L3	3	7	21APR2010
1001	Cycle 3	3	L4	4	7	21APR2010
1001	Cycle 4	4	L1	1	8	01JUL2010
1001	Cycle 4	4	L2	2	10	01JUL2010
1001	Cycle 4	4	L3	3	8	01JUL2010
1001	Cycle 4	4	L4	4	7	01JUL2010

ODS Style Template Used

The report is produced in two Output Delivery System (ODS) destinations, the Rich Text Format (RTF) destination, and the PDF destination. The Analysis style is used and specified prior to the PROC REPORT section in the statements:

```
ods rtf style=analysis file="{PATH\FILENAME}.rtf";
ods rtf style=analysis file="{PATH\FILENAME}.pdf";
```

Programs Used

One program, Ch4qc.sas, is used to create the Lesion Data Quality Report.

Implementation

COMPUTE Block Variables: DATA Step (Temporary) Versus REPORT (COLUMN Statement) Variables

Two types of variables are used in the COMPUTE blocks shown throughout this chapter. A brief description of each variable type's features relevant to this chapter will provide a basis for further understanding.

DATA step variables, or **temporary variables**, are created and used only in COMPUTE blocks. Some features specific to DATA step variables include:

- These variables are not reinitialized as new rows are processed. These temporary variables retain their values until changed by COMPUTE block code.
- DATA step variables cannot be used as report items in the PROC REPORT COLUMN statement.

REPORT variables, or **COLUMN statement variables** are referenced in the PROC REPORT COLUMN statement. These are report items, whether printed or not. REPORT variables can come from the input data set, or they can be created by a COMPUTE block (and they are DEFINEd as COMPUTED). Note, when reading through the chapter example:

- If an output data set is requested, even when the NOPRINT option is applied, the variable will be part of the output data available for data set processing.
- REPORT variables are reinitialized at every row; values are not retained.
- Display attributes of REPORT variables can be defined using CALL DEFINE statements.

Additional sources for detailed explanations on PROC REPORT variables and how they are processed can be found in the references at the end of this book, including Booth (2012), Carpenter (2007), and the *Base SAS® 9.3 Procedures Guide, 2nd Edition*.

The remainder of the chapter describes the specific steps implemented to develop the report shown in Figures 4.1 and 4.2.

ORDER by and Print Subject ID on Every Row with Greying Font

This report calls for ordering rows by Subject ID as well as printing the Subject ID on every report row. In addition, the subject ID should be shown in grey font after the first occurrence.

It is simple to order rows by Subject ID by applying the ORDER or GROUP option in the SUBJECT DEFINE statement. However, one of the results of ORDER/GROUP usage is that it prevents repeated values of the variable from being printed after the first occurrence. Since one of the report requirements is to print the Subject ID on every row, we need to find a work around.

The solution for obtaining ordered rows and printed SUBJECT values on every row is found by using a COMPUTE block to create a new subject variable (named ColSubj here) that can be DEFINEd as COMPUTED. Once we have two Subject ID variables, the ORDER option along with the NOPRINT option is associated with SUBJECT to maintain the report order, and the COMPUTED option is applied to ColSubj, the variable to be printed on each row.

The solution for conditionally applying black or grey font is also accomplished within the COMPUTE block. A temporary variable that holds each subject's record count is created, and a CALL DEFINE statement is used to apply the font style to the REPORT column version based on subject record count.

Program for Subject ID Display

```
** Get Lesion Data;
data lesion;
   set sasuser.Ch4Lesn;
run;

** ODS Specifications;
ods escapechar = "^";
options nodate nonumber orientation=landscape;
ods _all_ close;
ods rtf style=analysis file="C:\Users\User\My Documents\APR\Ch4.rtf";
```

Chapter 4: Lesion Data Quality Report: COMPUTE Blocks 103

```
ods pdf style=analysis file="C:\Users\User\My Documents\APR\Ch4.pdf";

** Produce the Report;

title "Potential Data Issues Report";

proc report data=lesion nowd split="|" center missing
   style(column)=[font_size=11 pt cellwidth=.9 in]
   style(header)=[font_size=11 pt background=white]
   style(summary)=[font_weight=bold]
   OUT=CH4DS;

columns subject ColSubj ❶ VisitN visit LesN LesC ScanDT SizeMM;


   define subject / order noprint; ❷
   define ColSubj / computed  "Subject ID"; ❸
   define VisitN  / order order=internal noprint;
   define visit   / order order=internal "Visit";
   define LesN    / analysis N noprint;
   define LesC    / order order=internal "Lesion ID"
                    style(column)=[cellwidth=1.4 in];
   define ScanDT  / order "Assessment|Date"  style(column)=[just=l
                    cellwidth=1.1 in];
   define SizeMM  / order "Lesion|Size (mm)" style(column)=[just=c
                    cellwidth=1 in];

** Initialize DATA Step variable before new subject;
   compute before subject;
      DtaSubCt=0; ❹
   endcomp;
```

```
** Retain Subject Number across Records in Output;
** Copy to DATA from Table variable and then copy back for rows where it
   would normally be blank;

  compute ColSubj / character length=16;  ❺
     if subject ne " " then DtaSubj = subject;  ❻
     ColSubj = DtaSubj;  ❼


** Increment Within Subject Counter Variable for Greying;  ❽
     if upcase(_BREAK_) = "SUBJECT" then DtaSubCt = 0;
     else if _BREAK_ = " " then DtaSubCt + 1;


** Apply Styles based on the Subject's Record Count;  ❾
     if DtaSubCt > 1 then
         call define("ColSubj","style","style={foreground=grey}");
     else call define("ColSubj","style","style={font_weight=bold}");
  endcomp;
```

```
run;

ods _all_ close;

title;

footnote;

ods html;
```

❶ ColSubj is added to the COLUMN statement.

❷ In the DEFINE statement for SUBJECT, the NOPRINT option is added. SUBJECT will now only be used for ordering rows, while ColSubj will be printed.

❸ In the DEFINE statement for ColSubj, the COMPUTED specification is applied and a column label is added.

❹ DtaSubCt, a DATA step variable that will be used to increment each subject's record count, is initialized to zero before each new subject ID.

❺ A new column variable, ColSubj is COMPUTED. It is declared as a character variable and the length is set. If "character" or "char" was not specified, PROC REPORT would assume that ColSubj is numeric.

Note that getting ColSubj to contain Subject ID on every row is **not** as simple as writing ColSubj = SUBJECT; many COMPUTE values would be blank due to use of the ORDER option for SUBJECT. Rather, a two-step process is used:

❻ The DATA step variable DtaSubj (rather than the column variable ColSubj) is set equal to SUBJECT. Because of a DATA step variable's automatic retain feature, DtaSubj will retain Subject.

❼ Because DATA step variables cannot be used as report items, ColSubj is then set equal to DtaSubj. ColSubj now contains a Subject ID for every row.

❽ The counter variable, DtaSubCt is incremented by 1 for each record within a subject ID where the _BREAK_ variable is blank (detail records).

Note that the temporary variables DtaSubCt and DtaSubj are manipulated within the COMPUTE ColSubj block - there is no "COMPUTE DtaSubCt" nor "COMPUTE DtaSubj" block.

❾ The grey font color is applied to the printed variable ColSubj after the subject's first record.

For reference, Table 4.3 shows a partial print of the PROC REPORT output data set created in the PROC REPORT statement (via "OUT=CH4DS"). Initially you may wonder why we had to make use of a DATA Step variable (DtaSubj) to retain the Subject ID values when SUBJECT appears on every row of the data set. The reason is, the available values for a COMPUTE block using an ORDER (or GROUP) variable are based on the printed output values (in this example the PDF and RTF reports), not the incoming nor output data set values. In other words, the repeated values of SUBJECT are no longer available in the PROC REPORT internal working file for the COMPUTE block. They are available, however, for rows written to the data set CH4DS.

Also note how the automatic variable _BREAK_ is either populated with the value "SUBJECT", or a blank. The rows for which _BREAK_ is blank represent detail rows (e.g. individual subject records).

The rows in which _BREAK_ contains "SUBJECT" values are summary rows created by the COMPUTE BEFORE SUBJECT block.

> Note that the summary lines contain each subject's lesion (LESN) count. This is because we specify the N statistic in the LESN DEFINE statement. Had we not specified a statistic for this ANALYSIS variable, the lesion ID's (e.g., 1, 2, 3, 4) would have been summed (the default statistic). Though lesion counts are not relevant here, they will be in a later step when we count lesions per visit.

Table 4.3 Partial Print of PROC REPORT Data Set CH4DS

SUBJECT	ColSubj	VisitN	VISIT	LesN	LesC	ScanDT	SizeMM	_BREAK_
1001	1001	.		45		.	.	SUBJECT
1001	1001	0	Screening	1	L1	10OCT2010	10	
1001	1001	0	Screening	1	L2	10OCT2010	12	
1001	1001	0	Screening	1	L3	10OCT2010	12	
1001	1001	0	Screening	1	L4	10OCT2010	11	
1001	1001	1	Cycle 1	1	L1	19DEC2010	9	
1001	1001	1	Cycle 1	1	L2	19DEC2010	10	
1001	1001	1	Cycle 1	1	L3	19DEC2010	10	
1001	1001	1	Cycle 1	1	L3	19DEC2010	11	
1001	1001	1	Cycle 1	1	L4	19DEC2010	10	
1001	1001	1	Cycle 1	1	L4	20DEC2010	10	
1001	1001	2	Cycle 2	1	L1	11FEB2010	8	
1001	1001	2	Cycle 2	1	L4	11FEB2010	9	
1001	1001	3	Cycle 3	1	L1	21APR2010	8	
1001	1001	3	Cycle 3	1	L2	21APR2010	7	
1001	1001	3	Cycle 3	1	L3	21APR2010	7	
1001	1001	3	Cycle 3	1	L4	21APR2010	7	
1001	1001	4	Cycle 4	1	L1	01JUL2010	8	
1001	1001	4	Cycle 4	1	L2	01JUL2010	10	
1001	1001	4	Cycle 4	1	L3	01JUL2010	8	
1001	1001	4	Cycle 4	1	L4	01JUL2010	7	
1001	1001	5	Cycle 5	1	L1	12AUG2010	9	
1001	1001	5	Cycle 5	1	L1	13AUG2010	8	

SUBJECT	ColSubj	VisitN	VISIT	LesN	LesC	ScanDT	SizeMM	_BREAK_
1001	1001	5	Cycle 5	1	L2	13AUG2010	7	
1001	1001	5	Cycle 5	1	L3	13AUG2010	8	
1001	1001	5	Cycle 5	1	L4	13AUG2010	9	
1001	1001	6	Cycle 6	1	L1	24SEP2010	8	
1001	1001	6	Cycle 6	1	L2	24SEP2010	7	
1001	1001	6	Cycle 6	1	L3	24SEP2010	8	
1001	1001	6	Cycle 6	1	L4	24SEP2010	9	
1001	1001	7	Cycle 7	1	L1	01NOV2010	12	
1001	1001	7	Cycle 7	1	L2	01NOV2010	.	
1001	1001	7	Cycle 7	1	L3	01NOV2010	7	
1001	1001	7	Cycle 7	1	L4	01NOV2010	8	
1001	1001	8	Cycle 8	1	L1	16DEC2010	7	
1001	1001	8	Cycle 8	1	L2	16DEC2010	5	
1001	1001	8	Cycle 8	1	L3	16DEC2010	7	
1001	1001	8	Cycle 8	1	L4	16DEC2010	5	
1001	1001	9	Cycle 9	1	L1	28JAN2011	21	
1001	1001	9	Cycle 9	1	L2	28JAN2011	7	
1001	1001	9	Cycle 9	1	L3	28JAN2011	5	
1001	1001	9	Cycle 9	1	L4	28JAN2011	5	
1001	1001	10	Cycle 10	1	L1	09MAR2011	4	
1001	1001	10	Cycle 10	1	L2	09MAR2011	8	
1001	1001	10	Cycle 10	1	L3	09MAR2011	3	
1001	1001	10	Cycle 10	1	L4	09MAR2011	9	
1002	1002	.		1		.	.	SUBJECT
1002	1002	0	Screening	1	L1	01JAN2011	6	
1003	1003	.		1		.	.	SUBJECT

Identify Potential Data Issues

Per the study protocol, each patient should have received one scan per lesion at each visit. Therefore, a check is put in place to indicate if patients have received more than one scan for a lesion at a particular visit. Checks are also put in place to assess if the number of lesions assessed at each visit matches the number of lesions assessed at the baseline (in this case, Screening) visit. In the case that a patient's lesion has completely disappeared, the lesion size is reported as zero millimeters.

Four data issue columns are added to the report to identify these potential data issues:

- Multiple dates per visit for a lesion
- Multiple sizes per visit for a lesion
- Greater number of lesions than assessed at baseline
- Fewer number of lesions than assessed at baseline

If any of these cases exist, a check mark is placed in the particular data issue column.

A subset of the case "Greater number of lesions than assessed at baseline" is the case in which a subject has no Screening visit, yet they have post-Screening visit(s). For this situation, an additional callout is made by concatenating the superscript "a" with the Subject's ID. The superscript "a" corresponds to the footnote "Subject has no Screening Visit."

Code for Displaying Potential Data Issues

The following PROC REPORT code is the same as that shown earlier, but with code added to identify subjects with no Screening visit and handle the new check-mark columns.

```
** Create Check Mark Formats; ❶
proc format;
  value check
    . , low - 1 = " "
    1 < - high = "^{unicode 2714}";
  value checkbi
    . , 0 = " "
    other = "^{unicode 2714}";
run;

** ODS Specifications;
ods escapechar = "^";
```

Chapter 4: Lesion Data Quality Report: COMPUTE Blocks 109

```sas
options nodate nonumber orientation =landscape;

ods _all_ close;

ods rtf style=analysis file="C:\Users\User\My Documents\APR\Ch4.rtf";

ods pdf style=analysis file="C:\Users\User\My Documents\APR\Ch4.pdf";

** Produce the Report;

title "Potential Data Issues Report";

** New Footnote to Identify Subjects with No Screening Visit;

footnote j=l "^S={indent=.75 in}^{super a}Subject has no Screening
Visit";

proc report data=lesion nowd split="|" center missing
   style(column)=[font_size=11 pt cellwidth=.9 in]
   style(header)=[font_size=11 pt background=white]
   style(summary)=[font_weight=bold]
   OUT=CH4DS;

** Add Four New Report Variables to COLUMN Statement;  ❷

   columns subject ColSubj VisitN visit LesN LesC ScanDT SizeMM
           ColDtCt ColSzDf ColMore ColFewer;

   define subject / order noprint;

   define ColSubj / computed "Subject ID";

   define VisitN  / order order=internal noprint;

   define visit   / order order=internal "Visit";

   define LesN    / analysis N noprint;

   define LesC    / order order=internal "Lesion ID"
                    style(column)=[cellwidth=1.4 in];

   define ScanDT  / order "Assessment|Date"   style(column)=[just=l
                    cellwidth=1.1 in];

   define SizeMM  / order "Lesion|Size (mm)"  style(column)=[just=c
```

```
                        cellwidth=1 in];
```

** Add Four Corresponding DEFINE Statements for the New Report Variables; ❸

```
  define ColDtCt  / computed "Multiple Dates Per Lesion ID"
format=check.
                      style(header column)=[background=yellow] center;
  define ColSzDf  / computed "Multiple Sizes Per Lesion ID"
format=checkbi.
                      style(header column)=[background=yellow] center;
  define ColMore  / computed "More Lesions" format=checkbi.
                      style(header column)=[background=powderblue]
center;
  define ColFewer / computed "Fewer Lesions" format=checkbi.
                      style(header column)=[background=powderblue]
center;
```

** Initialize DATA Step Variables Before New Subject and New Lesion ID; ❹

```
  compute before subject;
    DtaSubCt=0;
    DtaBlCt =0;
    DtaLesCt=0;
  endcomp;

  compute before LesC;
    DtaDtCt = 0;
    DtaFDt  = .;
    DtaSize = 0;
  endcomp;
```

** Populate DATA Step Variables that Count Number of Lesions Per Visit; ❺

```
  compute before visit;
```

```
   ** Insert Blank Line before each New Visit;
   line " ";


   **  Retain Baseline Lesion Count Across all Records;
   if visit = "Screening" then DtaBlCt = LesN.n;


   **  Get Lesion Count for each Visit Record;
   DtaLesCt = LesN.n;
 endcomp;

** Get summary record containing Total Lesion Count per Visit; ❻
 break after visit /summarize suppress;
 compute after visit;
    LesC= "Lesion Count = "||left(put(LesN.n,8.));  ❼
 endcomp;
```

```
** Retain Subject Number across Records in Output;
** Copy to DATA from Table variable and then copy back for rows
where it would normally be blank;
  compute ColSubj / character length=16;
     if subject ne " " then DtaSubj = subject;
        ColSubj = DtaSubj;
        ** NEW CODE TO ADD SUPERSCRIPT TO SUBJECT WITH NO SCREENING
           VISIT;
       if DtaBlCt = 0 and DtaLesCt > 0 then ColSubj =  strip(Dtasubj)
|| "^{super a}";

      ** Increment Within Subject Counter Variable for Greying;
   if upcase(_BREAK_) = "SUBJECT" then DtaSubCt = 0;
   else if _BREAK_ = " " then DtaSubCt + 1;

   ** Apply Styles based on the Subject's Record Count;
```

112 *PROC REPORT by Example: Techniques for Building Professional Reports Using SAS*

```
        if DtaSubCt > 1 or (DtaSubCt=1 and upcase(_BREAK_) = 'VISIT')
then
           call define("ColSubj","style","style={foreground=grey}");
        else call define("ColSubj","style","style={font_weight=bold}");
   endcomp;
```

```
** Count Scan Dates per Lesion within Visit; ❽

   compute ColDtCt;

      ** Increment the date count variable;

      if _BREAK_ = "LesN" then DtaDtCt = 0;

      else if _BREAK_ = " " then

         do;

            DtaDtCt + 1;


            ** Retain First Date for Lesion within Visit;

            if DtaDtCt = 1 then DtaFDt = ScanDt;

            ** Transfer DATA Step Variable Value to Report Variable;
            ** Multiple Dates Per Lesion ID;

            if DtaFDt ne ScanDt and Scandt ne . then ColDtCt=DtaDtCt;

         end;

   endcomp;

** In case of Multiple Scans (DtaDtCt > 1) Assess if Lesion Size is
   Different; ❾

   compute ColSzDf;

      ** Get Size Difference in lesion size from first scan within
visit;

      if SizeMM ne . and DtaDtCt = 1 then DtaSize = SizeMM;

      if DtaDtCt>1 then

         do;
```

```
            if nmiss(DtaSize,SizeMM)=0 then DtaSzDif = SizeMM-DtaSize;
         ** Copy DATA Step Variable to Report Variable;
         **Multiple Sizes Per Lesion ID;
         ColSzDf = DtaSzDif;
      end;
   endcomp;

   ** Column for cases of Lesion Count > Screening Visit; ❿
   compute ColMore;
      if upcase(_BREAK_) = "VISIT" and DtaLesCt-DtaBlCt > 0 then ColMore=1;
   endcomp;

   ** Column for cases of Lesion Count < Screening Visit; ⓫
   compute ColFewer;
      if upcase(_BREAK_) = "VISIT" and DtaLesCt-DtaBlCt < 0 then ColFewer=1;
   endcomp;

run;

ods _all_ close;

title;

footnote;

ods html;
```

❶ The check-mark formats are specified. "^{unicode 2714}" creates the heavy check mark symbol.

❷ The four new REPORT variables need to be added to the column statement.

❸ DEFINE statements are added for the four COMPUTEd variables. The appropriate check mark formats and background colors are applied.

❹ Creates additional DATA step variables. DtaSubCt, DtaBlCt, and DtaLesCt are initialized to zero before each new Subject ID. DtaDtCt, DtaFDt, and DtaSize are initialized before each new lesion ID (LesC) for a particular subject visit.

❺ We want to obtain lesion counts and carry them forward across records. Temporary variables are used for this purpose. DtaBLCt contains the Screening Visit lesion count, and DtaLesCt contains each visit's lesion count.

❻ A BREAK statement and corresponding COMPUTE block are employed to print the total lesion count summary row after each visit.

❼ For each summary row after VISIT, LesC will contain the text "Lesion Count =" followed by the actual lesion count.

❽ Creates COMPUTED column variable ColDtCt to contain the count of scan date records per lesion.

- The DATA step variable, DtaDtCnt is used to count scan date (ScanDT) records per lesion by incrementing DtaDtCnt by 1 each time a new record is encountered within a subject VISIT.
- The DATA step variable, DtaFDt retains the first scan date within a subject VISIT.
- In the case of multiple records, if the current scan date differs from the first scan date, the DATA step variable value is passed through to the report variable ColDtCt. A check mark format is applied to this column.

❾ Creates COMPUTED column variable ColSzDf to contain the difference between the current record's lesion size and the lesion size at the first visit date for scan dates in which DtaDtCt > 1 (records that are not the first date within the visit).

- The DATA step variable DtaSize is set to SizeMM (lesion size) at the first date within a visit (where DtaDtCnt=1) for a non-missing lesion size. A comparison of the current record's lesion size to the lesion size at the first visit date is obtained for scan dates.
- The DATA step variable value is passed through to the report variable ColSzDf.

❿ Creates COMPUTED column variable ColMore to flag cases of a visit having more lesions than were present at the Screening visit.

⓫ Creates COMPUTED column variable ColFewer to flag cases of a visit having fewer lesions than were present at the Screening visit.

Table 4.4 displays a partial print of the output data set and Figure 4.3 displays a partial portion of the final report. The shaded rows of the data set represent rows that also print in the final report. These include all detail records (i.e. where _BREAK_ = " ") and the VISIT Summary Lines.

We do not print the BEFORE SUBJECT, BEFORE VISIT, and BEFORE LesC summary rows in the final report, as these are only used to create interim DATA step variables.

Table 4.4 Partial PROC PRINT (Variables SUBJECT and VisitN Excluded Due to Size Restriction)

ColSubj	VISIT	LesN	LesC	ScanDT	SizeMM	ColDtCt	ColSzDf	ColMore	ColFewer	_BREAK_
1001		45		.	.	.	.	.	.	SUBJECT
1001	Screening	4		.	.	.	.	.	.	VISIT
1001	Screening	1	L1	.	.	.	.	.	.	LesC
1001	Screening	1	L1	10OCT2010	10	.	.	.	.	
1001	Screening	1	L2	.	.	.	.	.	.	LesC
1001	Screening	1	L2	10OCT2010	12	.	.	.	.	
1001	Screening	1	L3	.	.	.	.	.	.	LesC
1001	Screening	1	L3	10OCT2010	12	.	.	.	.	
1001	Screening	1	L4	.	.	.	.	.	.	LesC
1001	Screening	1	L4	10OCT2010	11	.	.	.	.	
1001	Screening	4	Lesion Count = 4	.	.	.	.	.	.	VISIT
1001	Cycle 1	6		.	.	.	.	.	.	VISIT
1001	Cycle 1	1	L1	.	.	.	.	.	.	LesC
1001	Cycle 1	1	L1	19DEC2010	9	.	.	.	.	
1001	Cycle 1	1	L2	.	.	.	.	.	.	LesC
1001	Cycle 1	1	L2	19DEC2010	10	.	.	.	.	
1001	Cycle 1	2	L3	.	.	.	.	.	.	LesC
1001	Cycle 1	1	L3	19DEC2010	10	.	.	.	.	
1001	Cycle 1	1	L3	19DEC2010	11	.	.	1	.	
1001	Cycle 1	2	L4	.	.	.	1	.	.	LesC
1001	Cycle 1	1	L4	19DEC2010	10	.	.	.	.	
1001	Cycle 1	1	L4	20DEC2010	10	2	0	.	.	
1001	Cycle 1	6	Lesion Count = 6	.	.	.	0	1	.	VISIT

Figure 4.3 Final Report

Potential Data Issues Report

Subject ID	Visit	Lesion ID	Assessment Date	Lesion Size (mm)	Multiple Dates Per Lesion ID	Multiple Sizes Per Lesion ID	More Lesions	Fewer Lesions
		Lesion Detail			Within Visit		Lesion Count Differs from Screening	
1001	Screening	L1	10OCT2010	10				
1001		L2	10OCT2010	12				
1001		L3	10OCT2010	12				
1001		L4	10OCT2010	11				
1001		Lesion Count = 4						
1001	Cycle 1	L1	19DEC2010	9				
1001		L2	19DEC2010	10				
1001		L3	19DEC2010	10				
1001				11		✓		
1001		L4	19DEC2010	10				
1001			20DEC2010	10	✓			
1001		Lesion Count = 6					✓	

Note that the DATA step variables (e.g., DtaBlCt and DtaLesCt) do not appear in the data set, nor in the printed output. They were only temporary variables stored in memory.

To help debug your code, you can temporarily copy the DATA Step variables to REPORT variables as shown in Figure 4.4. Both Table 4.5 and Figure 4.4 show variables relevant to the ColMore and ColFewer derivations. For interim debugging purposes, the NOPRINT option has been removed for LesN, DtaBlCt has been copied to a report variable named XColBlCt, and DtaLesCt has been copied to XColLesCt.

Table 4.5 Relevant Variables for ColMore and ColFewer Derivations

Description	Variable in Figure 4.4	Relevant Code
Count Lesions	LesN.n	COMPUTE BEFORE visit
Retain Baseline Lesion Count Across all Records	DtaBlCt (copied to XColBlCt)	if visit = "Screening" then DtaBlCt = LesN.n;
Get Lesion Count for each Visit Record	DtaLesCt (copied to XColLesCt)	DtaLesCt = LesN.n

Column for cases of Lesion Count > Screening Visit	ColMore	break after visit /summarize suppress;
		compute ColMore;
		   if upcase(_BREAK_) = "VISIT"
		   and DtaLesCt - DtaBlCt > 0
		   then ColMore = 1;
		endcomp;

Column for cases of Lesion Count < Screening Visit	ColFewer	break after visit /summarize suppress;
		compute ColFewer;
		   if upcase(_BREAK_) = "VISIT"
		   and DtaLesCt - DtaBlCt < 0
		   then ColFewer = 1;
		endcomp;

Figure 4.4 Interim Debugging Display

Potential Data Issues Report

Subject ID	Visit	LesN	Lesion ID	Assessment Date	Lesion Size (mm)	More Lesions	Fewer Lesions	Debug: Copy Temp Vars To Report Vars XColBlCt (DtaBlCt)	XColLesCt (DtaLesCt)
1001	Screening	1	L1	10OCT2010	10			4	4
1001		1	L2	10OCT2010	12			4	4
1001		1	L3	10OCT2010	12			4	4
1001		1	L4	10OCT2010	11			4	4
1001		4	Lesion Count = 4					4	4
1001	Cycle 1	1	L1	19DEC2010	9			4	6
1001		1	L2	19DEC2010	10			4	6
1001		1	L3	19DEC2010	10			4	6
1001		1			11			4	6
1001		1	L4	19DEC2010	10			4	6
1001		1		20DEC2010	10			4	6
1001		6	Lesion Count = 6			✔		4	6
1001	Cycle 2	1	L1	11FEB2010	8			4	2
1001		1	L4	11FEB2010	9			4	2
1001		2	Lesion Count = 2				✔	4	2

Both Table 4.6 and Figure 4.5 show variables relevant to the ColDtCt and ColSzDf derivations. For interim debugging purposes, DtaDtCt has been copied to a report variable named XColDtCt, DtaFDt has been copied to a report variable named XColFDt, and DtaSize has been copied to the report variable XColSize.

Table 4.6 Relevant Variables for ColDtCt and ColSzDf Derivations

Description	Variable in Figure 4.5	Relevant Code
Increment the date count variable	DtaDtCt (copied to XColDtCt)	if upcase(_BREAK_) = "LESC" then DtaDtCt = 0; else if _BREAK_ = " " then **do;** DtaDtCt + 1;
Retain First Date for Lesion within Visit	DtaFDt (copied to XColFDt)	if DtaDtCt = 1 then DtaFDt = ScanDt;

Chapter 4: Lesion Data Quality Report: COMPUTE Blocks   119

Description	Variable in Figure 4.5	Relevant Code
Transfer DATA Step Variable Value to Report Variable for Multiple Dates Per Lesion ID	ColDtCt (Check-Mark Column)	if DtaFDt ne ScanDt and ScanDt ne . then ColDtCt=DtaDtCt; end;
Get Lesion Size for First Record within Visit	DtaSize (Copied to XColSize)	if SizeMM ne . and DtaDtCt = 1 then DtaSize = SizeMM;
In case of Multiple Scans (DtaDtCt > 1) get Difference in Lesion Size from First Scan within Visit	ColSzDf (Check Mark Column)	if DtaDtCt>1 then do; if nmiss(DtaSize,SizeMM)=0 then DtaSzDif = SizeMM-DtaSize; ColSzDf = DtaSzDif; end;

Figure 4.5 Interim Debugging Display

Potential Data Issues Report

							Debug: Copy Temp Vars To Report Vars		
Subject ID	Visit	Lesion ID	Assessment Date (Scan Date)	Lesion Size (mm)	Multiple Dates Per Lesion ID	Multiple Sizes Per Lesion ID	XColDtCt (DtaDtCt)	XColFDt (DtaFDt)	XColSize (DtaSize)
1001	Screening	L1	10OCT2010	10			1	10OCT2010	10
1001		L2	10OCT2010	12			1	10OCT2010	12
1001		L3	10OCT2010	12			1	10OCT2010	12
1001		L4	10OCT2010	11			1	10OCT2010	11
1001							*1*	*10OCT2010*	*11*
1001	Cycle 1	L1	19DEC2010	9			1	19DEC2010	9
1001		L2	19DEC2010	10			1	19DEC2010	10
1001		L3	19DEC2010	10			1	19DEC2010	10
1001				11		✓	2	19DEC2010	10
1001		L4	19DEC2010	10			1	19DEC2010	10
1001			20DEC2010	10	✓		2	19DEC2010	10
1001							*2*	*19DEC2010*	*10*

Final Formatting: Create Spanning Headers

The final step to completing the report is to add formatting to the PROC REPORT output, including spanned headers. The code additions needed in the COLUMN statement are shown below.

```
columns
    ("^S={vjust=m bordertopcolor=white bordertopwidth=.02in}Lesion Detail"
        subject ColSubj VisitN visit LesN LesC ScanDT SizeMM
    ) ❶
    ("^S={vjust=m cellheight=.4 in}POTENTIAL DATA ISSUES"
        ("^S={background=yellow vjust=m} Within Visit" ColDtCt ColSzDf)
        ("^S={background=powderblue} Lesion Count|Differs from Screening"
        ColMore ColFewer)
    ); ❷
```

❶ The header "Lesion Detail" is spanned over the columns subject, ColSubj, VisitN, visit, LesN, LesC, ScanDT, and SizeMM. Because this creates an unwanted border over "Lesion Detail", the top border is hidden on the printed white page by changing the bordercolor to white and specifying width. The vjust=m portion of the STYLE (alias "S") override vertically centers the header within the cell.

❷ The header "POTENTIAL DATA ISSUES" spans ColDtCt, ColSzDf, ColMore, and ColFewer. Nested headers are created with the use of additional parentheses, where ColDtCt and ColSzDf get the header "Within Visit" and ColMore and ColFewer get the header "Lesion Count Differs from Screening."

Chapter 4 Summary

This chapter demonstrated differences between COMPUTE block DATA Step (Temporary) variables, and COMPUTE Block REPORT (COLUMN Statement) variables.

- Because DATA Step variables automatically retain their values (until changed by the programmer), we took advantage of DATA Step variables to count records and compare data values across visits.
- In the cases where a COMPUTE block variable needed to be part of the printed reports, we created a REPORT variable with a COMPUTE <REPORT VARIABLE> block.
- We also created new REPORT variables for the cases in which an incoming data set variable needed two usage types. For example, SUBJECT needed to be ORDERed and printed on every row. We couldn't simply create an alias, since aliases must have the same usage type.
- We created the new COMPUTED type subject variable.

- Check-marks were obtained by applying formats, and CALL DEFINEs were used to further highlight potential data issues with color.

Figure 4.1 Potential Data Issues Report

Potential Data Issues Report

Subject ID	Visit	Lesion ID	Assessment Date	Lesion Size (mm)	Multiple Dates Per Lesion ID	Multiple Sizes Per Lesion ID	More Lesions	Fewer Lesions
		Lesion Detail			Within Visit		Lesion Count Differs from Screening	
1001	Screening	L1	10OCT2010	10				
1001		L2	10OCT2010	12				
1001		L3	10OCT2010	12				
1001		L4	10OCT2010	11				
1001		*Lesion Count = 4*						
1001	Cycle 1	L1	19DEC2010	9				
1001		L2	19DEC2010	10				
1001		L3	19DEC2010	10				
1001				11		✓		
1001		L4	19DEC2010	10				
1001			20DEC2010	10	✓			
1001		*Lesion Count = 6*					✓	
1001	Cycle 2	L1	11FEB2010	8				
1001		L4	11FEB2010	9				
1001		*Lesion Count = 2*						✓
1001	Cycle 3	L1	21APR2010	8				
1001		L2	21APR2010	7				
1001		L3	21APR2010	7				
1001		L4	21APR2010	7				
1001		*Lesion Count = 4*						
1001	Cycle 4	L1	01JUL2010	8				

[a] Subject has no Screening Visit

Chapter 5: Multi-Sheet Workbook With Histograms—ExcelXP Tagsets Report

Introduction **124**

Example: Multi-Sheet Workbook Containing Heart Study Results **124**

Goals for Creating the Multi-Sheet Workbook **128**

    Key Steps 128

Source Data **129**

ODS Style Template Used **130**

Programs Used **134**

Implementation **134**

Create Formats and Informats **134**

    Code for Creating Formats and Informats 134

Obtain Counts and Percentages **137**

    Code for Obtaining Counts and Percentages 137

Producing the Workbook With PROC REPORT and ODS Tagset **141**

    Code for Opening, Closing, and Setting Initial Options for the ExcelXP Workbook 142

Producing the Specific Worksheets **144**

Code for Producing ByStatusCOL and ByStatusROW Worksheets 144

Code for Producing ByStatusALL Worksheet 149

Chapter 5 Summary **154**

Introduction

The SAS ExcelXP tagset, available with SAS 9.1 and later releases, allows SAS programmers to program a number of desirable Excel features directly into an Excel workbook. We can apply autofilters to columns so that the reader can select rows based on particular characteristics. We can provide ease of readability with features such as frozen headers for onscreen review and fit-to-page settings for printed reports. Other features include embedding titles and footnotes in the worksheets, page orientation, and scaling. The ExcelXP file created is an .XML file, which can be opened with Microsoft Excel 2002 or later.

Example: Multi-Sheet Workbook Containing Heart Study Results

A multi-sheet workbook is produced to report characteristics of alive and deceased patients that participated in a longitudinal heart study. Characteristics such as gender, age at key milestones, and a number of health status measures are presented in three different worksheets. Specifically, the worksheets include:

- Worksheet "ByStatusCOL": containing column percentages with histograms.
- Worksheet "ByStatusROW": containing row percentages with histograms.
- Worksheet: "ByStatusALL": containing a table of counts, overall percentages, column percentages, and row percentages.

Figure 5.1 displays the "ByStatusCOL" worksheet, which shows column percentages within alive versus deceased status. As an example, the gender percentages within the "Alive" column can be interpreted as the percentage of alive patients who are male versus female. Likewise, the gender percentages within the "Deceased" column can be interpreted as the percentage of deceased patients who were male versus female. The male and female percentages within each column sum to 100%.

Figure 5.1 Alive Status Column Percentages

Percentage of Alive/Deceased Patients having Various Characteristics (Column Percentages)

% of Alive			% of Deceased
	colspan	**Gender**	
	39%	Male	55%
	61%	Female	45%
		Age at Coronary Heart Disease (CHD) Diagnosis	
	83%	No CHD	55%
	1%	26 to 45	2%
	10%	46 to 65	26%
	7%	>65	18%
		Cholesterol Status	
	31%	Desirable	20%
	37%	Borderline	34%
	30%	High	42%
	3%	Missing	3%
		Blood Pressure Status	
	19%	Optimal	9%
	47%	Normal	32%
	34%	High	59%
		Weight Status	
	4%	Underweight	3%
	31%	Normal	23%
	65%	Overweight	73%
	0%	Missing	0%
		Smoking Status	
	62%	Non-smoker to Light (1-5)	54%
	11%	Moderate (6-15)	11%
	26%	Heavy (16-25) to Very Heavy (> 25)	34%
	0%	Missing	1%

Page 1 of 1 Pages

Figure 5.2 displays the "ByStatusROW" worksheet, which shows row percentages within various categories that are alive versus deceased. As an example, the gender percentages indicate the percentage of males that are alive versus deceased, and the percentage of females that are alive versus deceased. The alive and deceased percentages within each row sum to 100%.

Figure 5.2 ByStatusROW

Percentage of Patients who are Alive/Deceased by Various Characteristics
(Row Percentages)

Alive	% of ↴	Deceased
Gender		
53%	Male	47%
69%	Female	31%
Age at Coronary Heart Disease (CHD) Diagnosis		
71%	No CHD	29%
42%	26 to 45	58%
38%	46 to 65	62%
38%	>65	62%
Cholesterol Status		
71%	Desirable	29%
64%	Borderline	36%
53%	High	47%
55%	Missing	45%
Blood Pressure Status		
78%	Optimal	22%
70%	Normal	30%
48%	High	52%
Weight Status		
62%	Underweight	38%
69%	Normal	31%
59%	Overweight	41%
50%	Missing	50%
Smoking Status		
65%	Non-smoker to Light (1-5)	35%
63%	Moderate (6-15)	37%
55%	Heavy (16-25) to Very Heavy (> 25)	45%
44%	Missing	56%

Page 1 of 1 Pages

Figures 5.3 and 5.4 are displays of the printed report, and a snapshot of the Excel screen for the "ByStatusALL" worksheet. The screenshot illustrates how the worksheet appears with autofilters.

Figure 5.3 Print Version of "ByStatusALL"

Counts and Percentages by Patient Status

Category	Sub-Category	Alive Count	Alive Percent	Alive Column Percent	Alive Row Percent	Deceased Count	Deceased Percent	Deceased Column Percent	Deceased Row Percent
Gender	Male	1241	24%	39%	53%	1095	21%	55%	47%
Gender	Female	1977	38%	61%	69%	896	17%	45%	31%
Age at Coronary Heart Disease	No CHD	2663	51%	83%	71%	1097	21%	55%	29%
Age at Coronary Heart Disease	26 to 45	22	0%	1%	42%	31	1%	2%	58%
Age at Coronary Heart Disease	46 to 65	313	6%	10%	38%	508	10%	26%	62%
Age at Coronary Heart Disease	>65	220	4%	7%	38%	355	7%	18%	62%
Cholesterol Status	Desirable	998	19%	31%	71%	407	8%	20%	29%
Cholesterol Status	Borderline	1186	23%	37%	64%	675	13%	34%	36%
Cholesterol Status	High	951	18%	30%	53%	840	16%	42%	47%
Cholesterol Status	Missing	83	2%	3%	55%	69	1%	3%	45%
Blood Pressure Status	Optimal	626	12%	19%	78%	173	3%	9%	22%
Blood Pressure Status	Normal	1497	29%	47%	70%	646	12%	32%	30%
Blood Pressure Status	High	1095	21%	34%	48%	1172	22%	59%	52%
Weight Status	Underweight	113	2%	4%	62%	68	1%	3%	38%
Weight Status	Normal	1012	19%	31%	69%	460	9%	23%	31%
Weight Status	Overweight	2090	40%	65%	59%	1460	28%	73%	41%
Weight Status	Missing	3	0%	0%	50%	3	0%	0%	50%
Smoking Status	Non-smoker to Light (1-5)	2002	38%	62%	65%	1078	21%	54%	35%
Smoking Status	Moderate (6-15)	363	7%	11%	63%	213	4%	11%	37%
Smoking Status	Heavy (16-25) to Very Heavy (> 25)	837	16%	26%	55%	680	13%	34%	45%
Smoking Status	Missing	16	0%	0%	44%	20	0%	1%	56%

Page 1 of 1 Pages

In Figure 5.4, the autofilters can be seen in the "Category" and "Sub-Category" columns.

Figure 5.4 "ByStatusALL" Screen Shot with Auto Filters

	Category	Sub-Category	Alive				Deceased			
			Count	Percent	Column Percent	Row Percent	Count	Percent	Column Percent	Row Percent
6	Gender	Male	1241	24%	39%	53%	1095	21%	55%	47%
7	Gender	Female	1977	38%	61%	69%	896	17%	45%	31%
9	Age at Coronary Heart Disease	No CHD	2663	51%	83%	71%	1097	21%	55%	29%
10	Age at Coronary Heart Disease	26 to 45	22	0%	1%	42%	31	1%	2%	58%
11	Age at Coronary Heart Disease	46 to 65	313	6%	10%	38%	508	10%	26%	62%
12	Age at Coronary Heart Disease	>65	220	4%	7%	38%	355	7%	18%	62%
14	Cholesterol Status	Desirable	998	19%	31%	71%	407	8%	20%	29%
15	Cholesterol Status	Borderline	1186	23%	37%	64%	675	13%	34%	36%
16	Cholesterol Status	High	951	18%	30%	53%	840	16%	42%	47%
17	Cholesterol Status	Missing	83	2%	3%	55%	69	1%	3%	45%
19	Blood Pressure Status	Optimal	626	12%	19%	78%	173	3%	9%	22%
20	Blood Pressure Status	Normal	1497	29%	47%	70%	646	12%	32%	30%
21	Blood Pressure Status	High	1095	21%	34%	48%	1172	22%	59%	52%

Goals for Creating the Multi-Sheet Workbook

The goals for producing the .XML report include:

- Develop multiple worksheets within one Excel workbook.
- Provide column and row percentages with histograms.
- Provide one overall summary table.
- Incorporate Excel features such as autofilters, fittopage, repeat and freezing of headers, and the application of embedded and non-embedded titles and footnotes.
- Combine PROC REPORT capabilities (e.g., usage options for ordering rows, COMPUTE blocks with style options for histograms) and ExcelXP options.

Key Steps

Key steps include:

- Obtaining simple frequencies and percentages with PROC FREQ. The FREQ data sets are concatenated and used for later reporting.
- The ODS tagsets.ExcelXP destination is opened, and options that apply to all reports are specified.

- An additional ODS tagsets.ExcelXP statement is issued before each worksheet to identify sheet-specific options that should be employed.
- PROC REPORT code follows each worksheet ODS statement.
- After all three worksheets are created, the ods tagsets.ExcelXP destination is closed.

Source Data

The source data set for Chapter 5 is the SASHELP.HEART data set. Table 5.1 displays a partial print of the data, and Table 5.2 displays the contents of SASHELP.HEART.

Table 5.1 Partial Print of SASHELP.HEART Data

Obs	Status	DeathCause	AgeCHDdiag	Sex	AgeAtStart	Height	Weight	Diastolic	Systolic
1	Dead	Other		Female	29	62.50	140	78	124
2	Dead	Cancer		Female	41	59.75	194	92	144
3	Alive			Female	57	62.25	132	90	170
4	Alive			Female	39	65.75	158	80	128
5	Alive			Female	58	61.75	131	92	176
6	Alive			Female	36	64.75	136	80	112
7	Alive			Female	37	64.50	134	76	120
8	Alive			Female	42	67.75	162	96	138
9	Alive			Female	37	66.25	148	78	110
10	Alive			Female	45	64.00	147	74	120

SASHELP.HEART Data continued:

Obs	MRW	Smoking	AgeAtDeath	Cholesterol	Chol_Status	BP_Status	Weight_Status	Smoking_Status
1	121	0	55			Normal	Overweight	Non-smoker
2	183	0	57	181	Desirable	High	Overweight	Non-smoker
3	114	10		250	High	High	Overweight	Moderate (6-15)
4	123	0		242	High	Normal	Overweight	Non-smoker
5	117	0		196	Desirable	High	Overweight	Non-smoker
6	110	15		196	Desirable	Normal	Overweight	Moderate (6-15)
7	108	10		196	Desirable	Normal	Normal	Moderate (6-15)
8	119	1		200	Borderline	High	Overweight	Light (1-5)
9	112	15		192	Desirable	Optimal	Overweight	Moderate (6-15)
10	119	5		209	Borderline	Normal	Overweight	Light (1-5)

Table 5.2 SASHELP.HEART Contents

#	Variable	Type	Len	Label
1	Status	Char	5	
2	DeathCause	Char	26	Cause of Death
3	AgeCHDdiag	Num	8	Age CHD Diagnosed
4	Sex	Char	6	
5	AgeAtStart	Num	8	Age at Start
6	Height	Num	8	
7	Weight	Num	8	
8	Diastolic	Num	8	
9	Systolic	Num	8	
10	MRW	Num	8	Metropolitan Relative Weight
11	Smoking	Num	8	
12	AgeAtDeath	Num	8	Age at Death
13	Cholesterol	Num	8	
14	Chol_Status	Char	10	Cholesterol Status
15	BP_Status	Char	7	Blood Pressure Status
16	Weight_Status	Char	11	Weight Status
17	Smoking_Status	Char	17	Smoking Status

ODS Style Template Used

The SAS-supplied ODS Style Template STATISTICAL is modified and saved to a new template named STATISTICALX. We create a new ODS style template because we want to slightly alter the STATISTICAL template prior to producing the reports. The PARENT= statement specifies that STATISTICALX should inherit the style elements from STYLES.STATISTICAL.

We make the desired style element modifications for our new template via CLASS statements. Specifically, we change the fonts, modify the cell borders, change the background color of the NoteContent and titles, and modify the page margins.

```
** Modify Style Template;
proc template;
  define style styles.statisticalX;
  parent=styles.statistical;

    class fonts /
      "docfont"      = ("Helvetica", 10 pt)  /* Data in Table Cells */
      "EmphasisFont" = ("Tahoma", 12 pt, Bold) /* We will apply to
                                                  NoteContent */
      "headingfont"  = ("Tahoma", 18 pt, Bold) /* Table Column
                                                  Headings */
      "titlefont"    = ("Tahoma", 20 pt, Bold) /* System Title */
      ;

    class header /
      bordercolor=black
      bordertopwidth    =4 pt
      borderbottomwidth=4 pt
      borderleftwidth   =4 pt
      borderrightwidth =4 pt;

    class data /
      bordercolor=black
      bordertopwidth    =1 pt
      borderbottomwidth=1 pt
      borderleftwidth   =1 pt
      borderrightwidth =1 pt;

    ** COMPUTE Block Lines ;
    class NoteContent /
      font=fonts("EmphasisFont")
```

```
              borderleftwidth=1 pt
              borderrightwidth=1 pt
              backgroundcolor=white;
         class Body /
           bottommargin = .5 in
           topmargin    = .5 in
           rightmargin  = .5 in
           leftmargin   = .5 in;
         class systemtitle /
           backgroundcolor=white;
    end;
    run;
```

Figure 5.5 shows the ByStatusCOL worksheet based on the STATISTICAL Style Template before the PROC TEMPLATE changes. Figures 5.6 through 5.10 show the impact of one change at a time.

Figure 5.5 "ByStatusCOL" Without PROC TEMPLATE Changes

	Percentage of Alive/Deceased Patients having Various Characteristics (Column Percentages)	
% of Alive		% of Deceased
	Gender	
	39% Male	55%
	61% Female	45%

Figure 5.6 "ByStatusCOL" With Font Changes Added

Percentage of Alive/Deceased Patients having Various Characteristics (Column Percentages)		
% of Alive		**% of Deceased**
	Gender	
	39% Male	55%
	61% Female	45%

Figure 5.7 "ByStatusCOL" With Header (Specifically Border) Changes (See % of Alive, % of Deceased Row)

% of Alive		% of Deceased
	Gender	
39%	Male	55%
61%	Female	45%

Percentage of Alive/Deceased Patients having Various Characteristics (Column Percentages)

Figure 5.8 "ByStatusCOL" With Data (Specifically Border) Changes (See Male and Female Rows)

% of Alive		% of Deceased
	Gender	
39%	Male	55%
61%	Female	45%

Percentage of Alive/Deceased Patients having Various Characteristics (Column Percentages)

Figure 5.9 "ByStatusCOL" With NoteContent (Specifically Font, Border, and Background) Changes (See Gender Row)

% of Alive		% of Deceased
	Gender	
39%	Male	55%
61%	Female	45%

Percentage of Alive/Deceased Patients having Various Characteristics (Column Percentages)

Figure 5.10 "ByStatusCOL" With Systemtitle Changes (See Title Background Color)

Percentage of Alive/Deceased Patients having Various Characteristics (Column Percentages)

% of Alive		% of Deceased
	Gender	
39%	Male	55%
61%	Female	45%

Programs Used

CH5Tgxml.sas is the program used in this chapter.

Implementation

Create Formats and Informats

The formats and informats that will feed the procedures are created.

Code for Creating Formats and Informats

```
** Needed Formats and Informats; ❶
proc format;

  ** Formats;
  ** Collapse Age Categories;
  value age
    .      = "No CHD"
    26< - 45 = "26 to 45"
    45< - 65 = "46 to 65"
    65< - high = ">65";

  ** For Labeling the Report Rows;
  value $cat
    "SEX"       = "Gender"
    "DIAGAGE"   = "Age at Coronary Heart Disease (CHD) Diagnosis"
    "BP_STATUS" = "Blood Pressure Status"
    "WEIGHTCAT" = "Weight Status"
    "CHOLCAT"   = "Cholesterol Status"
    "SMOKECAT"  = "Smoking Status";
```

```
** For Applying Color to the Report;

value colora
  other=yellow
  .=white;

value colord
  other=powderblue
  .=white;

** For Ordering and Displaying ACROSS Variable;
** Add Split Character before Row*Percent to add a 3rd level(blank
   line) before these Header Labels;

value $name
  "1" = "Count"
  "2" = "Percent"
  "3" = "Column*Percent"
  "4" = "*Row*Percent";

** Collapse Smoker Categories;

value $smkcat
  "Non-smoker", "Light (1-5)"        = "Non-smoker to Light (1-5)"
  "Moderate (6-15)"                  = "Moderate (6-15)"
  "Heavy (16-25)", "Very Heavy (> 25)" =
                                       "Heavy (16-25) to Very Heavy (> 25)"
  " "                                = "Missing";

** ORDERED Version of ACROSS Variable;

value $statord
  "COUNT"   = "1"
  "PERCENT" = "2"
```

```
    "PCT_COL" = "3"
    "PCT_ROW" = "4";
```

** Informats - For Creating Numeric Variable Versions for Ordering;

```
  invalue age
    "No CHD"   = 0
    "26 to 45" = 1
    "46 to 65" = 2
    ">65"      = 3;

  invalue bp
     "Optimal" = 1
     "Normal"  = 2
     "High"    = 3
     "Missing" = 4;

  invalue chol
     "Desirable"  = 1
     "Borderline" = 2
     "High"       = 3
     "Missing"    = 4;

  invalue sex
    "Male"  =1
    "Female"=2;

  invalue smk
     "Non-smoker to Light (1-5)"              = 1
     "Moderate (6-15)"                        = 2
     "Heavy (16-25) to Very Heavy (> 25)" = 3
     "Missing"                                = 4;
```

```
    invalue wgt
      "Underweight" = 1
      "Normal"      = 2
      "Overweight"  = 3
      "Missing"     = 4;
    invalue varord
      "SEX"         = 1
      "DIAGAGE"     = 2
      "CHOLCAT"     = 4
      "BP_STATUS"   = 5
      "WEIGHTCAT"   = 6
      "SMOKECAT"    = 7;
run;
```

Obtain Counts and Percentages

The FREQ procedure is used to obtain counts and percentages by alive / deceased status for each of the categories to be reported (e.g., "Gender," "Age at Coronary Heart Disease (CHD) Diagnosis," etc.).

A macro is used to pass each FREQ variable through the procedure and obtain a different output data set for each variable. The data sets are then stacked so that each variable and its values are displayed in one column, as shown in Figures 5.1 through 5.3. The approach used to stack variables into columnar format is similar to the approach used in both Chapters 1 and 3.

The code for obtaining the counts and percentages is shown below.

Code for Obtaining Counts and Percentages

```
** Get data;
proc sort data=sashelp.heart out=heart;
   by sex;
```

```
run;

** Data Set Variable Recodes;
data heart;
   length diagage cholcat weightcat smokecat $60;
   set heart;
   ** CHD Age at Diagnosis;
   diagage=put(agechddiag,age.);
   smokecat = put(smoking_status,$smkcat.);
   if chol_status=" " then cholcat="Missing";
   else cholcat = chol_status;
   if weight_status=" " then weightcat="Missing";
   else weightcat = weight_status;
run;

** Frequency and Percentages Macro;
** Specify OUTPCT so output data set includes row, column, and table
   percentages;
%MACRO FREQ(freqvar=, fmt=);
   proc freq data=heart;
      tables &freqvar*status / missing outpct crosslist out=&freqvar;
   run;
   ** Create character string and add varname;
   data &freqvar;
      length &freqvar $60 varname $30;
      set &freqvar;
      varname= upcase("&freqvar");
      subctord=input(&freqvar,&fmt..);
```

```sas
   run;

   ** Transpose the freq Data;

   proc sort data=&freqvar;
      by varname &freqvar subctord status ;
   run;

   proc transpose data=&freqvar
        out=&freqvar(drop=_LABEL_;
      by varname &freqvar subctord;
      id status;
      var COUNT PERCENT pct_row pct_col;
   run;

   data &freqvar(drop=&freqvar);
      length subcat $60;
      set &freqvar;
      subcat=&freqvar;
   run;
%MEND freq;

%freq(freqvar=sex,fmt=sex)
%freq(freqvar=diagage,fmt=age)
%freq(freqvar=cholcat,fmt=chol)
%freq(freqvar=bp_status,fmt=bp)
%freq(freqvar=smokecat,fmt=smk)
%freq(freqvar=weightcat,fmt=wgt)

** Combine the FREQ Data Sets;
data status;
   set
      sex
```

```
            diagage
            cholcat
            bp_status
            smokecat
            weightcat;
    if _name_ in("COUNT","PERCENT","PCT_ROW","PCT_COL") then
       do;
          ** Put percentages in decimal format for Excel file;
          if _name_ ne "COUNT" then
             do;
                if alive ne . then alive=alive/100;
                if dead ne . then dead=dead/100;
             end;
        name=put(_name_,$statord.);
       end;
    varord=input(varname,varord.);
  run;
```

Table 5.3 displays a partial print of the combined patient data.

Table 5.3 Partial Print of Combined Patient Data

subcat	varname	subctord	_NAME_	Alive	Dead	name	varord
Female	SEX	2	COUNT	1977.00	896.00	1	1
Female	SEX	2	PERCENT	0.38	0.17	2	1
Female	SEX	2	PCT_ROW	0.69	0.31	4	1
Female	SEX	2	PCT_COL	0.61	0.45	3	1
Male	SEX	1	COUNT	1241.00	1095.00	1	1
Male	SEX	1	PERCENT	0.24	0.21	2	1
Male	SEX	1	PCT_ROW	0.53	0.47	4	1

subcat	varname	subctord	_NAME_	Alive	Dead	name	varord
Male	SEX	1	PCT_COL	0.39	0.55	3	1
26 to 45	DIAGAGE	1	COUNT	22.00	31.00	1	2
26 to 45	DIAGAGE	1	PERCENT	0.00	0.01	2	2
26 to 45	DIAGAGE	1	PCT_ROW	0.42	0.58	4	2
26 to 45	DIAGAGE	1	PCT_COL	0.01	0.02	3	2
46 to 65	DIAGAGE	2	COUNT	313.00	508.00	1	2
46 to 65	DIAGAGE	2	PERCENT	0.06	0.10	2	2
46 to 65	DIAGAGE	2	PCT_ROW	0.38	0.62	4	2
46 to 65	DIAGAGE	2	PCT_COL	0.10	0.26	3	2
>65	DIAGAGE	3	COUNT	220.00	355.00	1	2
>65	DIAGAGE	3	PERCENT	0.04	0.07	2	2
>65	DIAGAGE	3	PCT_ROW	0.38	0.62	4	2
>65	DIAGAGE	3	PCT_COL	0.07	0.18	3	2
No CHD	DIAGAGE	0	COUNT	2663.00	1097.00	1	2
No CHD	DIAGAGE	0	PERCENT	0.51	0.21	2	2
No CHD	DIAGAGE	0	PCT_ROW	0.71	0.29	4	2
No CHD	DIAGAGE	0	PCT_COL	0.83	0.55	3	2

The data is now in the format needed for PROC REPORT.

Producing the Workbook With PROC REPORT and ODS Tagset

As with other ODS destinations, we sandwich the needed report procedure code within ODS opening and closing statements. Also, as with other destinations, the opening statement specifies the path, file, and ODS Style template used.

Immediately after enabling the destination, we set desired ODS TAGSET options in another ODS TAGSETS Statement. These options will apply to the workbooks that follow unless we add another ODS statement that changes the options before a specific worksheet.

Code for Opening, Closing, and Setting Initial Options for the ExcelXP Workbook

```
ods escapechar="^";

ods _all_ close;

ods tagsets.ExcelXP path="c:\Users\User\My Documents\APR\"
file="FHStudy.xml" style=statisticalX;   ❶

** Set some "global" tagset options that affect all/most worksheets;   ❷

ods tagsets.ExcelXP options(embedded_titles='yes'
                            orientation='Landscape'
                            frozen_headers='yes'
                            row_repeat='1-3'
                            fittopage='yes'
                            pages_fitwidth='1'
                            pages_fitheight='100'
                         Print_Footer='&LPage &amp;P of &amp;N Pages'
                            autofit_height='yes');

**    {PROC REPORT CODE GOES HERE...};

ods tagsets.ExcelXP close;   ❸

ods html;
```

❶ Specify the XML output file path, file name, and style template to be used.

❷ For this Excel workbook, we would like several user-specified options to apply to all or most spreadsheets. The tagset options remain in effect until they are turned off or until other values are specified. Therefore, options that we would like applied only to specific worksheets are implemented directly before that worksheet. Note, if no options are specified, the SAS ExcelXP default options are applied. The SAS default settings for TAGSETS.EXCELXP options, along with a history of changes, can be found at http://support.sas.com/rnd/base/ods/odsmarkup/excelxp_help.html.

Some of the options impact the printed report, some impact our view of Excel onscreen, and some impact both. For this report, when we view the Excel Workbook (onscreen) we would like the following options:

- Autofilters to be able to filter specific rows (for ByStatusALL worksheet)
- Frozen headers, so when we page down we can still see the column headers
- Embedded titles (so we can see the title in the worksheet rather than having to go to print preview). We do not embed the footnotes, so Page x of y Pages will show only in the print preview or on the printed reports.

For the printed report, we employ these options to allow more space on the horizontal page:

- A landscaped report
- A sheet fitting to one page wide
- Page x of y Pages footnotes

The options we apply to all of the worksheets are shown in Table 5.4.

Table 5.4 Options to Apply to All Worksheets

Option	Result
embedded_titles='yes'	Puts title as a row in the worksheet (versus in the header section seen only in print and print preview)
orientation='Landscape'	Orientation is set to 'Landscape' (versus default 'Portrait')
frozen_headers='yes'	Freezes column headers so when user scrolls down the column headers can still be seen
row_repeat='1-3	The first three rows are repeated when user scrolls down (the frozen header)
fittopage='yes'	Fits to a page
pages_fitwidth='1'	Specifies 1 page across
pages_fitheight='100'	Specifies up to 100 pages down
Print_Footer='&LPage &P of &N Pages'	Left justify Page x of y Pages footnote
autofit_height='yes'	Autofit row height

❸ The ODS Close statement closes and releases the XML file so that it can be opened with Excel.

144 *PROC REPORT by Example: Techniques for Building Professional Reports Using SAS*

Producing the Specific Worksheets

The first two worksheets use virtually the same program, with the exception of the data subset (where _name_ is "PCT_COL" or "PCT_ROW") and the sheet name assigned within the Excel workbook. A macro is used for these two worksheets.

The third worksheet presents the data in a different manner, i.e., as a table versus percentages with bars. Therefore, the PROC REPORT code for the summary table is run separately.

Code for Producing ByStatusCOL and ByStatusROW Worksheets

```
%MACRO PRRPT1( RepNum=   /* Report Number used for %IF-%THEN logic */
              ,PageTit= /* Title to appear at top of page */
              ,sheet=   /* Specify Sheet Name */
              ,pct=     /* Specify either column or row percentages */
             ); ❶

title "&PageTit"; ❷

ods tagsets.ExcelXP options(sheet_name="&sheet"); ❸

proc report data=status(where=(_name_="&pct"))
     nowd missing split="*"; ❹

   /** Column Specifications; **/ ❺
   column varord varname subctord

   /** Column Headers Specific to Column Percentages Worksheet **/
   %IF &REPNUM=1 %THEN %DO;
     alive
     ("% of Alive"  abar alive=Alive2)
     subcat
     ("% of Deceased" Dead dbar);
   %END;
```

```
/** Column Headers Specific to Row Percentages Worksheet **/

%IF &REPNUM=2 %THEN %DO;

   alive

   ("Alive"  abar alive=Alive2)

   ("% of ^{unicode 21B4}" subcat)

   ("Deceased" Dead dbar);

%END;

/** DEFINE Specifications; **/ ❻

define varord   /  group order=internal noprint;

define varname  /  group order=formatted noprint;

define alive    /  noprint;


define subctord / group order=internal noprint;

define subcat   / "" group order=formatted
                   style(column)=[tagattr="format:@"
                   cellwidth=2.25 in just=c];

/** Percents and Histograms; **/

   define Alive2    / "" style(column)=[cellwidth=.5 in just=r
                        tagattr="format:0%"];


   define abar      / "" style(column)=[cellwidth=4.4 in
                        font_size=5 pt just=r cellpadding=0
                        vjust=middle foreground=blue
                        font_weight=bold tagattr="format:@"];


    define Dead     / "" style(column)=[cellwidth=.5 in
                        tagattr="format:0%"];
```

146 *PROC REPORT by Example: Techniques for Building Professional Reports Using SAS*

```
define dbar    / ""  style(column)=[cellwidth=4.4 in
                         font_size=5 pt cellpadding=0 vjust=middle
                         foreground=blue font_weight=bold
                         tagattr="format:@"];
```

```
/** Initialize Temporary Variables; **/ ❼
compute before subcat;
   abarsize=0;
   dbarsize=0;
endcomp;

/** Create Row Headers; **/ ❽
compute before varname;
   line "";
   line varname $cat.;
endcomp;

/** HISTOGRAM; **/
/** "Histogram-Like" Bars (Alive Status); **/ ❾
  compute abar / char length=1500;
     if alive.sum=. then abar=" ";
     else do;
        abarsize=round(alive.sum*100);
        if abarsize>0 then
                    abar=strip(repeat("^{unicode 25AE}", abarsize));
     end;
  endcomp;

/** "Histogram-Like" Bars (Deceased Status); **/ ❿
```

```
    compute dbar / char length=1500;
      if dead.sum=. then dbar=" ";
      else do;
        dbarsize=round(dead.sum*100);
        if dbarsize>0 then
                       dbar=strip(repeat("^{unicode 25AE}", dbarsize));
      end;
    endcomp;
run;
quit;
%MEND PRRPT1;
```

/** BYSTATUSCOL AND BYSTATUSROW MACRO CALLS; **/ ❶

```
%PRRPT1(RepNum=1
       ,PageTit=%str(Percentage of Alive/Deceased Patients having
Various Characteristics^n(Column Percentages))
       ,sheet=ByStatusCOL
       ,pct=PCT_COL);

%PRRPT1(RepNum=2
       ,PageTit=%str(Percentage of Patients who are Alive/Deceased by
Various Characteristics^n(Row Percentages))
       ,sheet=ByStatusROW
       ,pct=PCT_ROW);
```

❶ The parameters for the PRRPT1 macro include:
- **REPNUM** – to distinguish Column Percentage and Row Percentage reports
- **PAGETIT** – to specify the title we would like to appear at the top of the page
- **SHEET** – to specify a worksheet name
- **PCT** – to subset data on either column or row percentages

❷ The embedded page title will depend on what is specified for &PAGETIT in the macro call.

❸ An additional ODS TAGSET option is added to for these two worksheets:
- A unique sheet name is applied (with SHEET=&sheet, based on the macro call)

❹ Each report is run on the subset of data specified in the macro call (where _name_ is "PCT_COL" or "PCT_ROW").

❺ Headers will be based on the particular report, as designated by the %DO - %END loop dependent on the &REPNUM macro variable value.

❻ The NOPRINT option applies to columns for which the purpose is ordering rows or performing other interim steps.

The ABAR and DBAR statements apply to the "Alive" and "Deceased" histograms, respectively. These are not created with a graph, but rather use the technique of repeating characters shared by Pete Lund in his paper, "You Did That Report in SAS®!?: The Power of the ODS PDF Destination" (See the "References" section of this book).

- **ABAR** is the bar for "Alive" percentages (the variable is derived in a following COMPUTE block).
- **DBAR** is the bar for "Deceased" percentages and uses the same method as ABAR.
- **TAGATTR** style attributes are used to send Microsoft formats from SAS to Excel. We use tagattr="format:0%" to ensure the percentages appear in the form x% (e.g., .39 displays as 39%). We use tagattr="format:@" to tell Excel that this should be treated as a text column.

❼ The bar sizes are initialized to zero prior to each new category so the value of the last bar of the previous category is not retained.

❽ Row headers are produced via LINE statements.

❾ Bars are only desired for non-missing percentages. ABARSIZE, a temporary variable captures the rounded percentage of ALIVE after converting the decimal to a percentage. Per COMPUTE block rules, the analysis variables used to derive the computed variables are referenced by their compound names (*VARIABLE.statistic*) such as ALIVE.sum.

The character variable ABAR then uses the REPEAT function to repeat the rectangle (❶) ABARSIZE times. For example, the rectangle is repeated 38 times for the percentage 38%. We create the rectangle with the code "^{unicode 25AE}".

Note that ABAR is declared as a character variable and the length is set to 1500. The length is set to handle the largest possible value, which is 1500 characters: a possible ABARSIZE of 100 (because a bar could have a maximum of 100%) times the number of characters in "^{unicode 25AE}", which is 15.

❿ The bars representing percentages for deceased patients (DBAR) are created in the same manner.

Code for Producing ByStatusALL Worksheet

The final worksheet contains a summary table for counts, total percentages, column percentages, and row percentages. We have all of these statistics from the FREQ procedure, so the PROC REPORT procedure is simply used to display these.

```
** Overall Summary Worksheet;
title "Counts and Percentages by Patient Status";
ods tagsets.ExcelXP options(sheet_name="StatusByALL"
                            autofilter="1-2"
absolute_column_width='13,13,4,4,4,4,4,4,4,4'
                            title_footnote_width='10');  ❶
proc report data=STATUS nowd missing split="*"
   style(header)=[font_size=14 pt font_face=Georgia font_weight=bold];
   column varord varname ("Category" colvname)
       subctord ("Sub-Category" subcat)
     ( ("Alive" alive) ("Deceased" dead)), NAME;  ❷
   define varord   / group order=internal noprint;
   define varname  / group order=formatted noprint;
   define colvname / "" computed format=$cat.
                       style(header)=[just=l];  ❸
   define subctord / group order=internal noprint;
   define subcat   / "" group;
   define name     / across order=internal "" format=$name.
                       style(header)=[font_size=10 pt];  ❹
   define Alive    / "";
   define Dead     / "";
   ** Computed Description Column that will Print on Every Row;
```

```
        compute colvname / char length= 30;
          if _break_ =" " and varname ne " " then dsvname=varname;
          colvname = dsvname;
        endcomp;

   ** Apply Background Colors and TAGATTR Formats;  ❺
   compute name;
      call define('_c6_',"style","style=[background=colora.]");
      call define('_c7_',"style","style=[background=colora.
                   tagattr='format:0%']");
      call define('_c8_',"style","style=[background=colora.
                   tagattr='format:0%']");
      call define('_c9_',"style","style=[background=colora.
                   tagattr='format:0%']");
      call define('_c10_',"style","style=[background=colord.]");
      call define('_c11_',"style","style=[background=colord.
                   tagattr='format:0%']");
      call define('_c12_',"style","style=[background=colord.
                   tagattr='format:0%']");
      call define('_c13_',"style","style=[background=colord.
                   tagattr='format:0%']");
   endcomp;

   compute before varname;
      line " ";
   endcomp;
run;
quit;
```

```
** CLOSE THE TAGSETS.EXCELXP Destination;
ods tagsets.ExcelXP close; ❻
ods html;
```

❶ For the StatusByALL worksheet, we want four tagset specifications added to the options list. These include the worksheet name, columns to which autofilters should be applied, the absolute column widths, and the number of columns over which the titles and footnotes should span for this specific worksheet. The StatusByALL options are listed just prior to this worksheet creation.

Option	Function
sheet_name="StatusByALL"	Name the worksheet
autofilter='1-2'	Apply auto filter to columns 1 & 2
absolute_column_width='13,13,4,4,4,4,4,4,4,4');	The first two columns will be a width of 13. The remaining columns will be a width of 4. See additional explanation below.
title_footnote_width='10'	Specifies that the titles and footnotes are allowed to span 10 columns. See additional explanation below.

Absolute_Column_Width

Vince DelGobbo explains in his paper "Creating Stylish Multi-Sheet Microsoft Excel Workbooks the Easy Way with SAS®(2011)" that the ExcelXP tagset uses the following formula to compute the **approximate** column width:

ColumnWidth = WIDTH_POINTS * ABSOLUTE_COLUMN_WIDTH * WIDTH_FUDGE

If we do not specify values for these tagset options, the value for that option is determined for us.

- WIDTH_POINTS refers to the font metrics
- ABSOLUTE_COLUMN_WIDTH is determined by the following rules:
    - For numeric variables, the absolute_column_width is the largest number of characters in any cell of the column, including the heading.
    - For character variables, the absolute_column_width is the larger of either length of the character variable, or the largest number of characters in a cell in the column.

- WIDTH_FUDGE is a multiplier that can be used to make small adjustments to all column widths (the default value is 0.75).

We specified the absolute_column_width value for each column in a comma-separated list of values. Note, if we had just specified absolute_column_width='13', assuming we did not apply column width overrides in PROC REPORT, all 10 columns would have an absolute_column_width of '13.'

Title_Footnote_Width

It is worth noting that without the title_footnote_width='10' specification, the embedded title "Counts and Percentages by Patient Status" would not have been centered on the page, as shown in Figure 5.11. We needed to specify the number of columns so that our title would be centered across all ten columns.

Figure 5.11 Title Placement Without the Title_Footnote_Width Option Specification

Counts and Percentages by Patient Status

		Alive				Deceased			
Category	Sub-Category	Count	Percent	Column Percent	Row Percent	Count	Percent	Column Percent	Row Percent
Gender	Male	1241	24%	39%	53%	1095	21%	55%	47%
Gender	Female	1977	38%	61%	69%	896	17%	45%	31%

> For more explanation on TAGSETS.EXCELXP tagset options, see DelGobbo (2011) listed in the references section of this book.

❷ Note that NAME, which contains Count and Percentage items, is DEFINEd as an across variable. As an across variable, PROC REPORT creates a column for each value of NAME (Count, Percent, Column Percent, Row Percent).

In a COLUMN statement that contains an across variable, a comma is needed between the across variable and the variables that are to be displayed beneath or above the across variable. Had we specified NAME before the comma, as in

```
NAME,(("Alive" alive) ("Deceased" dead))
```

we would arrive at the result in Figure 5.12. Note that to prevent wrapping of the word "Deceased," the header font size was reduced, and absolute column widths were modified.

Figure 5.12

Counts and Percentages by Patient Status

Category	Sub-Category	Count Alive	Count Deceased	Percent Alive	Percent Deceased	Column Percent Alive	Column Percent Deceased	Row Percent Alive	Row Percent Deceased
Gender	Male	1241	109500%	24%	21%	0.3856433	55%	53%	47%
Gender	Female	1977	89600%	38%	17%	0.6143567	45%	69%	31%

Rather, we place the variable NAME after the comma as in

    (("Alive" alive) ("Deceased" dead)), NAME

so the NAME values (Count, Percent, Colum Percent, Row Percent) are displayed beneath the "Alive" and "Deceased" variables, as shown in Figure 5.13.

Figure 5.13

Counts and Percentages by Patient Status

Category	Sub-Category	Alive Count	Alive Percent	Alive Column Percent	Alive Row Percent	Deceased Count	Deceased Percent	Deceased Column Percent	Deceased Row Percent
Gender	Male	1241	24%	39%	53%	1095	21%	55%	47%
Gender	Female	1977	38%	61%	69%	896	17%	45%	31%

Note that regardless of whether we choose NAME to span above or beneath Alive and Deceased status, the two variables (ALIVE and DEAD) need to be enclosed in parentheses.

❸ Because we are applying autofilters to the first two columns, we want the value of these columns to appear on every detail record. As shown in Chapter 4, ORDER variables print only the first occurrence of each value. In order to get the Category (VARNAME) to print on every row of its subset items, we use a COMPUTE block to create a COMPUTEd version of the Category variable (COLVAR).

❹ A format ($NAME) is applied to label the headers of the across variable (where "1" = "Count", "2" = "Percent", "3" = "Column Percent", and "4" = "Row Percent"). The ORDER of rows is specified as ORDER=Internal to preserve the unformatted 1, 2, 3, 4 order of the column display.

❺ Call DEFINE statements are used to fill in the ALIVE and DEAD column values with different colors. The actual ALIVE versus DEAD columns no longer exist because they are nested under an across variable. However, their corresponding columns can be accessed with their absolute column number variable names, which are _C6_ though _C13_.

Chapter 5 Summary

The ability to create reports in Excel multi-sheet workbooks allows the report recipients a number of options not available with other ODS destinations, such as RTF and PDF.

- For example, the reader may have multiple reports in the same workbook, each contained in its own worksheet. Subsets of the data can be selected with the use of autofilters. The familiar Excel user can sort and group the reports once they are in the Excel file.
- In addition to applying autofilters, this chapter demonstrated how to incorporate other ExcelXP features, such as freezing headers, embedding titles in the worksheet, displaying Page x of y Pages as a non-embedded footer, setting absolute column widths, fitting documents to a page, and providing repeated headers.
- TAGATTR style attributes were used to send Microsoft formats from SAS to Excel. Specifically, we used these to format data such as text and percentages within the Excel file data cell.
- PROC REPORT features were used to create histograms within the Excel report columns.

Chapter 6: Using the ACROSS Option to Create a Weekly Sales Report

Introduction **156**

Example: Weekly Sales Report **156**

Goals for Creating a Weekly Sales Report **158**
    Key Steps 158

Source Data **158**

ODS Style Template Used **160**

Programs Used **160**

Implementation: Creating the ODS Style Template **160**
    Proc Template Code 160

Obtain Calendar Grid and Merge With Sales **162**

Produce the Report **166**
    Code for Producing the Report 167

Place Holders for Data Not Yet Available **177**

Chapter 6 Summary **179**

Introduction

Some reports are better displayed once the data is transposed. While we could use PROC TRANSPOSE, PROC REPORT provides the advantage of allowing us to simultaneously transpose the data and summarize rows. This ability is provided with PROC REPORT's ACROSS and GROUP usage options. The variable we want transposed becomes the ACROSS variable, and the rows to be consolidated are the GROUP variables. Once the ACROSS option is applied, the variable's values become the columns of the report.

Example: Weekly Sales Report

A weekly report is provided to summarize Potato Chip sales from each Sunday to Monday period. The report is in a calendar format, with each weekday and its corresponding date across the top of the table, and product types displayed as the rows. A summary line is displayed above each table that shows the period reported (week start to week end), the total sales for that week, and the average sales amount per day.

Figure 6.1 displays the first page of the report.

Figure 6.1 Partial Print of a Weekly Sales Report

Week Start - End: 29DEC2002 - 04JAN2003   Total Sales: $515.44   Average/Day: $73.63

Product	Sun 29DEC	Mon 30DEC	Tue 31DEC	Wed 01JAN	Thu 02JAN	Fri 03JAN	Sat 04JAN	Total
Baked potato chips	$0	$0	$0	$0	$0	$0	$0	$0
Barbeque potato chips	$19.37	$23.84	$8.94	$17.88	$11.92	$8.94	$20.86	$111.75
Classic potato chips	$31.68	$40.59	$112.86	$21.78	$14.85	$35.64	$67.32	$324.72
Ruffled potato chips	$13.41	$11.92	$8.94	$5.96	$13.41	$8.94	$16.39	$78.97
Salt and vinegar potato chips	$0	$0	$0	$0	$0	$0	$0	$0
WOW potato chips	$0	$0	$0	$0	$0	$0	$0	$0
Total	$64.46	$76.35	$130.74	$45.62	$40.18	$53.52	$104.57	$515.44

Week Start - End: 05JAN2003 - 11JAN2003   Total Sales: $422.87   Average/Day: $60.41

Product	Sun 05JAN	Mon 06JAN	Tue 07JAN	Wed 08JAN	Thu 09JAN	Fri 10JAN	Sat 11JAN	Total
Baked potato chips	$0	$0	$0	$0	$0	$0	$0	$0
Barbeque potato chips	$17.88	$11.92	$10.43	$10.43	$14.90	$17.88	$16.39	$99.83
Classic potato chips	$31.68	$34.65	$32.67	$20.79	$26.73	$26.73	$60.39	$233.64
Ruffled potato chips	$13.41	$11.92	$10.43	$11.92	$17.88	$11.92	$11.92	$89.40
Salt and vinegar potato chips	$0	$0	$0	$0	$0	$0	$0	$0
WOW potato chips	$0	$0	$0	$0	$0	$0	$0	$0
Total	$62.97	$58.49	$53.53	$43.14	$59.51	$56.53	$88.70	$422.87

Goals for Creating a Weekly Sales Report

Creating a weekly sales report requires developing a dynamic (data driven) template. Other than specifying the desired reporting period, the programmer can simply run the report weekly and the tables will populate as new data arrives.

Key Steps

Preprocessing steps include creating a data set containing the desired reporting dates and merging this with the actual sales data.

PROC REPORT is then used to:

- Transpose dates across the top
- Consolidate rows
- Create total column and rows
- Display text above each table containing summary information

Source Data

The source data is a subset of the SASHELP data set "SNACKS". Table 6.1 displays a partial PROC PRINT of the data set. Table 6.2 displays the contents of SNACKS.

Table 6.1 Partial Print of SNACKS Data

QtySold	Price	Date	Product
13	1.49	29DEC2002	Barbeque potato chips
16	1.49	30DEC2002	Barbeque potato chips
6	1.49	31DEC2002	Barbeque potato chips
12	1.49	01JAN2003	Barbeque potato chips
8	1.49	02JAN2003	Barbeque potato chips
6	1.49	03JAN2003	Barbeque potato chips
14	1.49	04JAN2003	Barbeque potato chips
12	1.49	05JAN2003	Barbeque potato chips
8	1.49	06JAN2003	Barbeque potato chips
7	1.49	07JAN2003	Barbeque potato chips
7	1.49	08JAN2003	Barbeque potato chips
10	1.49	09JAN2003	Barbeque potato chips
12	1.49	10JAN2003	Barbeque potato chips
11	1.49	11JAN2003	Barbeque potato chips
11	1.49	12JAN2003	Barbeque potato chips
6	1.49	13JAN2003	Barbeque potato chips
9	1.49	14JAN2003	Barbeque potato chips
6	1.49	15JAN2003	Barbeque potato chips
17	1.49	16JAN2003	Barbeque potato chips

Table 6.2 Contents of SNACKS Data

#	Variable	Type	Len	Format	Label
1	QtySold	Num	8		Quantity sold
2	Price	Num	8		Retail price of product
3	Date	Num	8	DATE9.	Date of sale
4	Product	Char	40		Product name

ODS Style Template Used

A custom ODS style template is created. As a start, the ODS style template "SASWEB" is used as the parent style. The modified template is named "SASWEBR."

Programs Used

The program for this chapter is titled CH6CAL.SAS.

Implementation: Creating the ODS Style Template

We make slight modifications to the SAS-supplied "SASWEB" ODS style template and name the new template "SASWEBR".

Proc Template Code

```
** Make Modifications to SASWEB ODS Style Template;
proc template;
    define style styles.saswebr;
    parent=styles.sasweb;
  class fonts /
    "docfont"      = ("Garamond", 12 pt) /* Data in Table Columns */
    "EmphasisFont" = ("Garamond",12.5 pt, Bold) /* Line Statements */
    "headingfont"  = ("Garamond", 12 pt, Bold) ; /* Column Headers */
  class table /
    borderwidth=2 pt
    cellspacing=1.97 pt;
  class body /
    bottommargin = .25in
    topmargin    = .25in
    rightmargin  = .75in
```

Chapter 6: Using the ACROSS Option to Create a Weekly Sales Report   161

```
      leftmargin   = .75in;

   ** Apply Emphasis Font to Line Statements;

   class NoteContent /

      font=fonts("EmphasisFont");

end;

run;
```

Figure 6.2 shows the PROC REPORT output using the SASWEB template, or, prior to making PROC TEMPLATE modifications.

Figure 6.2 PROC REPORT Output Prior to Making PROC TEMPLATE Modifications

Week Start - End: 29DEC2002 - 04JAN2003		Total Sales: $515.44		Average/Day: $73.63				
Product	Sun 29DEC	Mon 30DEC	Tue 31DEC	Wed 01JAN	Thu 02JAN	Fri 03JAN	Sat 04JAN	Total
Baked potato chips	$0	$0	$0	$0	$0	$0	$0	$0
Barbeque potato chips	$19.37	$23.84	$8.94	$17.88	$11.92	$8.94	$20.86	$111.75
Classic potato chips	$31.68	$40.59	$112.86	$21.78	$14.85	$35.64	$67.32	$324.72
Ruffled potato chips	$13.41	$11.92	$8.94	$5.96	$13.41	$8.94	$16.39	$78.97
Salt and vinegar potato chips	$0	$0	$0	$0	$0	$0	$0	$0
WOW potato chips	$0	$0	$0	$0	$0	$0	$0	$0
Total	$64.46	$76.35	$130.74	$45.62	$40.18	$53.52	$104.57	$515.44

Figure 6.3 shows the PROC REPORT output using the SASWEBR template, or, after making PROC TEMPLATE modifications.

Figure 6.3 PROC REPORT Output After Making PROC TEMPLATE Modifications

Week Start - End: 29DEC2002 - 04JAN2003			Total Sales: $515.44		Average/Day: $73.63			
Product	Sun 29DEC	Mon 30DEC	Tue 31DEC	Wed 01JAN	Thu 02JAN	Fri 03JAN	Sat 04JAN	Total
Baked potato chips	$0	$0	$0	$0	$0	$0	$0	$0
Barbeque potato chips	$19.37	$23.84	$8.94	$17.88	$11.92	$8.94	$20.86	$111.75
Classic potato chips	$31.68	$40.59	$112.86	$21.78	$14.85	$35.64	$67.32	$324.72
Ruffled potato chips	$13.41	$11.92	$8.94	$5.96	$13.41	$8.94	$16.39	$78.97
Salt and vinegar potato chips	$0	$0	$0	$0	$0	$0	$0	$0
WOW potato chips	$0	$0	$0	$0	$0	$0	$0	$0
Total	$64.46	$76.35	$130.74	$45.62	$40.18	$53.52	$104.57	$515.44

Fonts are modified for the Table, Header, and NoteContent portions of the report. The "NoteContent" modifications allow us to obtain the desired font for the COMPUTE block lines (the text above the table) and summary rows ("Total" rows).

The table borders are made wider with the borderwidth and cellspacing style attributes, providing greater distinction between tables and data cells than occurs with the default SASWEB ODS Style Template.

- The borderwidth (see table frame) is widened to 2 pt width.
- The space between cells in the inner table is increased to 1.97 pt width (kept slightly smaller than the white left and right borderwidths of the Day/Date headers so the grey background color does not overtake the white).

Obtain Calendar Grid and Merge With Sales

After the style template is developed, we need to create placeholder dates for the desired reporting period. For this example, the reports will cover December 29, 2002 through April 5, 2003. Once the start and end dates of the reporting period are specified (via macro variables), a CALDATES data set is created to contain a record for each date of the period. CALDATES is then merged with the sales data set (SNACKS).

```
** Identify Period to be Reported;  ❶

%let startdt='29Dec2002'D;

%let enddt  ='05Apr2003'D;
```

```
** Subset Source Data; ❷

data sales;
  set sashelp.snacks(keep=Product Date QtySold Price
                    where=(&startdt <= date <= &enddt and
                           find(product,'potato','i')>0));
run;

** Create Calendar Data Set; ❸

data caldates;
  do date = &startdt to &enddt by 1;
    format date wkstart wkend date9.;
    chday = put(date,weekdate3.);
    wkstart=intnx('week',date,0);
    wkend  =intnx('week',date,1)-1;
    output;
  end;
run;

** Merge Sales with Caldates; ❹

proc sort data=sales;
  by date;
run;

proc sort data=caldates;
  by date;
run;

data sales2;
  length wksten $60;
  merge sales
        caldates;
```

164 *PROC REPORT by Example: Techniques for Building Professional Reports Using SAS*

```
       by date;

    ** Derive WKSTEN and SALEAMT Variables;

       wksten =
          catx(' - ',strip(put(wkstart,date9.)),strip(put(wkend,date9.)));


       if nmiss(QtySold,price)=0 then saleamt = QtySold * price;

    run;
```

- ❶ Macro variables are created to specify start and end dates for the report. In this case, the reporting period is from December 29, 2002 to April 5, 2003.
- ❷ The SAS-supplied data set sashelp.snacks is subset on dates between Dec 29, 2002 through April 05, 2003 and snacks containing the word "potato." The "i" argument tells SAS to ignore case when searching for the string "potato".
- ❸ Via a DO loop, the following variables are derived and a record is output for each date:
  - DATE – Dates will range from our Start Date to our End Date, as specified in the DO loop with the macro variables &STARTDT and &ENDDT. Although we may not yet have sales data all the way through &ENDDT, we create the template that will be populated when data becomes available.
  - CHDAY – The three-letter abbreviation for the day that the date falls on (Sun, Mon, Tue, Wed…)
  - WKSTART – Each week's Start Date (Start 'day' is always Sunday).
  - WKEND – Each week's End Date (End 'day' is always Saturday).
  - The INTNX function is used to determine Week Start and End Dates. The default WEEK Start interval for INTNX is Sunday, so no shift in days is needed.
    - ○ Using the date of Tuesday, December 31, 2002 as an example, wkstart = intnx('week','31dec2002'd,0);
      - – translates into: "Determine the date of the start of the week (Sunday) that is 0 weeks from the week of December 31, 2002. The result is Sunday, December 29, 2002."
    - ○ Using the date of Tuesday, December 31, 2002, as an example, wkend = intnx('week','31dec2002'd,1) - 1;
      - – translates into: "Determine the date of the start of the week (Sunday) that is 1 week after the week of December 31, 2002, minus 1 day. The result is Saturday, January 4, 2003."

Table 6.3 displays an example of the calendar data.

Table 6.3 Partial Print of CALENDAR Data

Date	wkstart	wkend	chday
29DEC2002	29DEC2002	04JAN2003	Sun
30DEC2002	29DEC2002	04JAN2003	Mon
31DEC2002	29DEC2002	04JAN2003	Tue
01JAN2003	29DEC2002	04JAN2003	Wed
02JAN2003	29DEC2002	04JAN2003	Thu
03JAN2003	29DEC2002	04JAN2003	Fri
04JAN2003	29DEC2002	04JAN2003	Sat
05JAN2003	05JAN2003	11JAN2003	Sun
06JAN2003	05JAN2003	11JAN2003	Mon
07JAN2003	05JAN2003	11JAN2003	Tue
08JAN2003	05JAN2003	11JAN2003	Wed
09JAN2003	05JAN2003	11JAN2003	Thu
10JAN2003	05JAN2003	11JAN2003	Fri
11JAN2003	05JAN2003	11JAN2003	Sat
12JAN2003	12JAN2003	18JAN2003	Sun
13JAN2003	12JAN2003	18JAN2003	Mon
14JAN2003	12JAN2003	18JAN2003	Tue
15JAN2003	12JAN2003	18JAN2003	Wed
16JAN2003	12JAN2003	18JAN2003	Thu
17JAN2003	12JAN2003	18JAN2003	Fri
18JAN2003	12JAN2003	18JAN2003	Sat

❹ Sales data and Calendar data are merged by DATE. Two new variables are derived. WKSTEN, contains the header to be place above each week, for example, "29DEC2002 - 04JAN2003." SALEAMT multiplies QtySold times Price to arrive at a total sales amount for each record.

Table 6.4 displays a partial print of the merged data set.

166 *PROC REPORT by Example: Techniques for Building Professional Reports Using SAS*

Table 6.4 Partial Print of Combined CALENDAR and SALES Data

Date	wkstart	wkend	chday	wksten	Product	saleamt	QtySold	Price
29DEC2002	29DEC2002	04JAN2003	Sun	29DEC2002 - 04JAN2003	Baked potato chips	0.00	0	1.99
29DEC2002	29DEC2002	04JAN2003	Sun	29DEC2002 - 04JAN2003	Barbeque potato chips	19.37	13	1.49
29DEC2002	29DEC2002	04JAN2003	Sun	29DEC2002 - 04JAN2003	Classic potato chips	31.68	32	0.99
29DEC2002	29DEC2002	04JAN2003	Sun	29DEC2002 - 04JAN2003	Ruffled potato chips	13.41	9	1.49
29DEC2002	29DEC2002	04JAN2003	Sun	29DEC2002 - 04JAN2003	Salt and vinegar potato chips	0.00	0	2.49
29DEC2002	29DEC2002	04JAN2003	Sun	29DEC2002 - 04JAN2003	WOW potato chips	0.00	0	2.99
30DEC2002	29DEC2002	04JAN2003	Mon	29DEC2002 - 04JAN2003	Baked potato chips	0.00	0	1.99
30DEC2002	29DEC2002	04JAN2003	Mon	29DEC2002 - 04JAN2003	Barbeque potato chips	23.84	16	1.49
30DEC2002	29DEC2002	04JAN2003	Mon	29DEC2002 - 04JAN2003	Classic potato chips	40.59	41	0.99
30DEC2002	29DEC2002	04JAN2003	Mon	29DEC2002 - 04JAN2003	Ruffled potato chips	11.92	8	1.49
30DEC2002	29DEC2002	04JAN2003	Mon	29DEC2002 - 04JAN2003	Salt and vinegar potato chips	0.00	0	2.49
30DEC2002	29DEC2002	04JAN2003	Mon	29DEC2002 - 04JAN2003	WOW potato chips	0.00	0	2.99
31DEC2002	29DEC2002	04JAN2003	Tue	29DEC2002 - 04JAN2003	Baked potato chips	0.00	0	1.99
31DEC2002	29DEC2002	04JAN2003	Tue	29DEC2002 - 04JAN2003	Barbeque potato chips	8.94	6	1.49
31DEC2002	29DEC2002	04JAN2003	Tue	29DEC2002 - 04JAN2003	Classic potato chips	112.86	114	0.99
31DEC2002	29DEC2002	04JAN2003	Tue	29DEC2002 - 04JAN2003	Ruffled potato chips	8.94	6	1.49

Produce the Report

Some highlights of what is accomplished via the REPORT procedure include:

- Picture formats are applied to style the column header appearance (e.g. Sun 29DEC) and the sale amounts in the table cells (e.g., $0 rather than $0.00, "*" rather than a blank).

Chapter 6: Using the ACROSS Option to Create a Weekly Sales Report

- The report is run BY WKSTART (each week's Start Date).
- DATE is defined as an ACROSS variable.
- Total rows are obtained with a BREAK AFTER each new WKSTART.
- A Total column is derived as a COMPUTEd variable.
- A summary line displayed above each page consists of a text variable that contains concatenated information from both report and temporary variables.

Code for Producing the Report

```
** Create Dollar and Date Picture Formats;
proc format;
  picture saleamt (round)
       .           = "*"
       0           = "$0" (noedit)
       0< - high = "000,000,009.00" (prefix="$");  ❶
  picture datefmt (default=100)
     low-high    ='%a %0d%b' (datatype=date);  ❷
run;

** Macro Variables for COMPUTE Block Array to Align Numbers and
   Asterisks in Cells;
%let _c4_  = _c4_;
%let _c5_  = _c5_;
%let _c6_  = _c6_;
%let _c7_  = _c7_;
%let _c8_  = _c8_;
%let _c9_  = _c9_;
%let _c10_ = _c10_;

** Specify ODS Information;
ods escapechar='^';
```

168 *PROC REPORT by Example: Techniques for Building Professional Reports Using SAS*

```
    options nonumber nobyline nodate orientation=landscape;  ❸

ods _all_ close;

ods rtf file='C:\Users\User\My
Documents\SAS\BOOK\Programs\Ch6Across.rtf' style=saswebr STARTPAGE=NO
bodytitle;  ❹

** REPORT Procedure;

proc report data=sales2 nowd nocenter split='|' missing
  out=chkdat
  style(column)=[cellwidth=.7 in];


  by wkstart;  ❺

  column wkstart=wkstart2 ("Product" product) wksten date, saleamt
         ("Total" total);  ❻

  ** DEFINE Specifications;  ❼

  define wksten    / group ' ' noprint;

  define wkstart2 / group ' ' noprint;

  define date      / " "  across format=datefmt. order=internal
                     style(header)=[borderleftwidth=2 pt
                                    borderleftcolor=white
                                    borderrightwidth=2 pt
                                    borderrightcolor=white];

  define saleamt  / " "  analysis format=saleamt.;

  define product  / group ' ' style(column)=[cellwidth=2 in];

  define total    / ""  computed format=saleamt.
                    style(column)=[just=d];

  ** COMPUTE BLOCKS;

  ** Create Column Variable TOTAL;  ❽
```

```
compute total;
   if product ne ' ' then
     do;
       ** Determine number of non-missing days with sales data;
       numday = n(_c4_,_c5_,_c6_,_c7_,_c8_,_c9_,_c10_);
       ** Get total if at least one day has sales;
       if numday>=1
         then total = sum(_c4_,_c5_, _c6_,_c7_,_c8_,_c9_,_c10_);
     end;
   ** If all data is missing hide the asterisks by applying white
      font color;
      if total=. then
         call define(_row_,'style','style={foreground=white}');
endcomp;

** Obtain Total Rows;
break after wkstart   / summarize;  ❾

** Initialize Temporary Variable for Later Use; ❿
compute BEFORE wkstart;
   dstotal=total;
endcomp;


** Summarize Sales Data Above Page (Since STARTPAGE=NO, multiple
   REPORT pages appear on one RTF page); ⓫
compute before _page_ /style={just=l};
   length text $120;
   if dstotal ne . then
      do;
```

```
               text= 'Week Start - End: ' || strip(wksten)
                    || '     Total Sales: '
                    || strip(put(dstotal,saleamt.))
                    || '     Average/Day: '
                    || strip(put(dstotal/numday,saleamt.));

          ** Additional Note for Weeks with Some Missing Data;
          if 1 <= numday < 7 then text = strip(text)
                                        || "^n* = Data Not Yet Available";
       end;

       ** Above Page Text for Weeks with All Missing Data;
       else
          do;
             text= 'Week Start - End: ' || strip(wksten)
                  || '     Total Sales:          '
                  || "Data Not Yet Available";
          end;

       line ' ' ;
       line text $120.;
       line ' ' ;
    endcomp;

    ** Fill in Product Column for Total Row and Conditionally Format;  ⓬
    compute product /char length=30;
       if product ne ' ' then product=product;
       else if _break_='wkstart' then
          do;
             product='Total';
             call define(_row_,'style','style={font_weight=bold}');
```

```
      end;
  endcomp;


** Decimal Align Numbers and Center Align Asterisks; ⓑ
  compute date;
    array colname (7) $ ("&_c4_" "&_c5_" "&_c6_" "&_c7_" "&_c8_"
                         "&_c9_" "&_c10_");
    array coln    (7)    _c4_    _c5_    _c6_    _c7_    _c8_
                         _c9_    _c10_;
      do i=1 to dim(colname);
        if coln(i) > .z then
          call define(colname(i),'style','style={just=d}');
        else call define(colname(i),'style','style={just=c}');
      end;
  endcomp;
run;

ods rtf close;
ods html;
```

Picture formats are used to create templates for printing dollar amounts and dates.

Two types of pictures are applied for formatting the dollar amounts.

- Digit selectors - these are numeric characters (0 through 9) that correspond to positions for numeric variables. A nonzero versus a zero digit selector basically tells SAS whether leading zeros are applied.

    o Nonzero digit selectors print leading zeros. We use 9 as the nonzero digit selector in the SALEAMT format.

    o Zero digit selectors do not print leading zeros.

- Message characters – these are nonnumeric characters that print as specified in the picture.

❶ For the dollar amounts, we want:
- Values rounded. The ROUND option rounds the sales amount to the nearest decimal value given in the format specification, before formatting. This prevents truncation of the values.
- Missing values printed as "*". The "*" is a "message character" and the format prints as it appears in the picture format.
- "$0.00" printed as "$0". The NOEDIT option tells SAS to treat $0 as a message character and the format should print as it appears in this assignment. Without the NOEDIT option, numbers are treated as digit selectors (digit placeholders) and "$0.00" values would display as blanks in the report.
- Values greater than 0 are assigned the picture "000,000,009.00". The PREFIX= statement adds a US dollar sign before the sales amount. We make the format picture wide enough to contain the sales amounts and the "$" prefix.
- The 0s and 9s are the digit selectors and these are important in designating the desired level of leading zeros. We do not want leading zeros in this report.
  ○ A helpful reminder about digit selectors is offered in a paper by Andrew Karp, *"Getting in to the Picture (Format)"* who sources Pete Lund with the saying "nines print zeros and zeros print blanks."

Table 6.5 demonstrates how the amounts 19.37 and 5.96 would display after various picture formats are applied.

Table 6.5 Application of Picture Formats

Picture Format		Original: 19.37	Original: 5.96
009.99	Displays number as ->	19.37	5.96
099.99	Displays number as ->	19.37	05.96
999.99	Displays number as ->	019.37	005.96

❷ For the date format, we want dates to appear in a form such as "Sun 05JAN". The length for the format is set with DEFAULT= option. The (DATATYPE=DATE) is necessary to create a date format.
- %a specifies an abbreviated (3-letter) day (Mon, Tue, Wed).
- %d specifies the day of the month, and the 0 before the d adds a leading zero to numbers < 10.
- %b specifies the abbreviated month name (DEC, JAN, FEB).

Chapter 6: Using the ACROSS Option to Create a Weekly Sales Report   173

❸   The NOBYLINE option is specified because we will create custom headers within each BY table.

❹   The ODS style template SASWEBR is selected. STARTPAGE=NO prevents each BY WKSTART (Week Start) table from being printed on a separate page. Note that we have no titles or footnotes for this report. The bodytitle option, which specifies to put titles and footnotes directly above or below a table, is used here simply to create more space on the page for the tables.

❺   The BY statement gives us separate tables for each week rather than one very large table.

❻   Because WKSTART is used as a BY variable, an alias for WKSTART (named WKSTART2) is created so we can GROUP the variable and then use it for BREAK AFTER and COMPUTE BEFORE statements. If we did not create the alias, SAS would automatically create an alias and add log NOTEs.

SALEAMT follows the comma after the ACROSS variable and it is these values that are reported in the cells under the DATE values. If multiple variables were to be nested under DATE, these would be enclosed in a parentheses immediately following the comma. For example, if we wanted to nest TOTAL as well as SALEAMT, the code would read

```
column wkstart=wkstart2 ("Product" product) wksten
       date, (saleamt total);
```

This is not what we want. In this example, TOTAL is a standalone column (i.e. not nested). The parentheses are applied only to create the header "Total" for the TOTAL column.

```
column wkstart=wkstart2 ("Product" product) wksten
       date, saleamt ("Total" total);
```

❼   DEFINE SECTION

Since we want the sales to be summed, all of the character variables are defined as GROUP variables. The SALEAMT column is defined as ANALYSIS, for which SUM is the default statistic. The NOPRINT option is applied to variables used only for the purpose of grouping, breaking, or computing. Table 6.5 provides more explanation on each DEFINE statement.

Table 6.5 Descriptions of DEFINE Statements

Statement	Use
define wksten / group ' ' noprint;	Variable used in the compute before _page_ line statement
define wkstart2 / group ' ' noprint;	Alias created for ability to break and compute before/after WKSTART
define date / " " across format=datefmt. order=internal style(header)=[borderleftwidth=2 pt borderleftcolor=white borderrightwidth=2 pt borderrightcolor=white];	DATE is our across variable. Each value of DATE within the BY period will have its own column. The picture format DATEFMT is applied to achieve the desired headers. The ORDER is specified as internal because we want the dates displayed in chronological order. We want the left and right borders for the DATE portion of the header to be white. Therefore, header style overrides are applied in the DATE DEFINE statement.
define saleamt / " " analysis format=saleamt. style(column)=[just=d];	This is our Sales amount variable to be summed and printed according to the SALEAMT picture format
define product / group ' ' style(column)=[cellwidth=2 in];	The Product column to be GROUPED and printed
define total / "" computed format=saleamt. style(column)=[just=d];	The Computed Total column sums across days displayed at the rightmost of the report

Note that as an ACROSS variable, DATE will be transposed by PROC REPORT so that individual date values become columns. In the case of ACROSS variables, the column names take the form _Cn_, where n is the absolute column number. Our seven date/day values become _C4_ through _C10_, for columns 4 through 10. These are the column names that need to be referenced in any compute block.

Table 6.6 shows a partial print of the PROC REPORT output data set. **Note that the BY variable (wkstart) is not counted in the column numbers in the output data set (see gray row in Table 6.6).**

Table 6.6 Partial Print (Observations and Variables) of PROC REPORT Data Set

BY VARIABLE	Column 1	Column 2	Column 3	Column 4 (Sun 29DEC)	Column 5 (Mon 30DEC)	Column 6 (Tue 31DEC)
wkstart	**wkstart2**	**Product**	**wksten**	**_C4_**	**_C5_**	**_C6_**
29DEC2002	29DEC2002	Total		64.46	76.35	130.74
29DEC2002	29DEC2002	Baked potato chips	29DEC2002 - 04JAN2003	0.00	0.00	0.00
29DEC2002	29DEC2002	Baked potato chips	29DEC2002 - 04JAN2003	0.00	0.00	0.00
29DEC2002	29DEC2002	Barbeque potato chips	29DEC2002 - 04JAN2003	19.37	23.84	8.94
29DEC2002	29DEC2002	Classic potato chips	29DEC2002 - 04JAN2003	31.68	40.59	112.86
29DEC2002	29DEC2002	Ruffled potato chips	29DEC2002 - 04JAN2003	13.41	11.92	8.94
29DEC2002	29DEC2002	Salt and vinegar potato chips	29DEC2002 - 04JAN2003	0.00	0.00	0.00
29DEC2002	29DEC2002	WOW potato chips	29DEC2002 - 04JAN2003	0.00	0.00	0.00

❽ The COMPUTED column TOTAL, containing the weekly sales sum is created. A temporary variable, NUMDAY, is created to obtain the number of days with non-missing sales data. A TOTAL is obtained if at least one date (NUMDAY >= 1) in the row has sales. For cases where a week has no data, i.e. where TOTAL = ., a white foreground is applied to turn the asterisks white so they do not show in the table.

❾ The BREAK after WKSTART along with the summarize option creates a TOTAL row for each weekly reporting period.

❿ COMPUTE BEFORE WKSTART initializes the temporary variable DSTOTAL to equal TOTAL before each new week start. The DSTOTAL value will be carried forward due to temporary variables' automatic retain property. The DSTOTAL information is used for the summary line above each table.

⓫ The COMPUTE BEFORE _PAGE_ block produces the summary information above each page. Recall that multiple REPORT pages are placed on one page of the report because we requested STARTPAGE=NO in the ODS RTF statement.

We use the DSTOTAL variable we created in the COMPUTE BEFORE WKSTART block, and the NUMDAY variable created in the COMPUTE TOTAL block to determine what text will display above each page.

- If there is a sales total for that week, the temporary variable TEXT will contain the start and end date information along with Total and Average Sales. For example:

"Week Start - End: 29DEC2002 04JAN2003   Total Sales: $515.44   Average/Day: $73.63"

- If a given week has some missing sales, an asterisk will appear in the missing data cell and the text "* = Data Not Yet Available" will become an additional line of header text using our designated escape character with the newline function (^n). (See Figure 6.4 first report)
- If a given week has no sales data yet, TEXT will contain only the start and end date information, along with the note "Data Not Yet Available." (See Figure 6.4 second report)

The temporary variables DSTOTAL and NUMDAY are used to populate the Total Sales and Average/Day amounts, where Average/Day = DSTOTAL/NUMDAY. The picture format SALEAMT is applied to these character strings.

⓬ The report variable PRODUCT already exists, and we want to keep the existing values for the detail rows. The purpose of this compute block is to assign a value of 'Total' for the summary row that we obtained with the summarize option. This assignment is made specifically "if _break_='wkstart' ." Table 6.7 displays a partial PRINT of the PROC REPORT data set.

Table 6.7 Partial Print of PROC REPORT Data Set Showing _BREAK_ Column and Total Rows

Wkstart	wkstart2	Product	wksten	_Cn_ Columns...	total	_BREAK_
29DEC2002	29DEC2002	Total			515.44	wkstart
29DEC2002	29DEC2002	Baked potato chips	29DEC2002 - 04JAN2003		0.00	_PAGE_
29DEC2002	29DEC2002	Baked potato chips	29DEC2002 - 04JAN2003		0.00	
29DEC2002	29DEC2002	Barbeque potato chips	29DEC2002 - 04JAN2003		111.75	
29DEC2002	29DEC2002	Classic potato chips	29DEC2002 - 04JAN2003		324.72	
29DEC2002	29DEC2002	Ruffled potato chips	29DEC2002 - 04JAN2003		78.97	
29DEC2002	29DEC2002	Salt and vinegar potato chips	29DEC2002 - 04JAN2003		0.00	
29DEC2002	29DEC2002	WOW potato chips	29DEC2002 - 04JAN2003		0.00	
29DEC2002	29DEC2002	Total			515.44	wkstart
05JAN2003	05JAN2003	Total			62.97	wkstart
05JAN2003	05JAN2003		05JAN2003 - 11JAN2003		.	_PAGE_

❸ We want decimal alignment for numbers, and center alignment for asterisks. CALL DEFINE statements are used to conditionally apply the desired alignment. An array is used so we do not need to write separate IF-THEN-ELSE sections for each DATE column (_C4_ through _C10_).

Place Holders for Data Not Yet Available

The program is set up to put an asterisk ("*") in empty cells, if one or more of the date columns in a given week can be populated with sales data. If there is no data for the entire week, the note "Data Not Yet Available" is simply printed above the table rather than populating the entire table with asterisks.

178 *PROC REPORT by Example: Techniques for Building Professional Reports Using SAS*

Figure 6.4 displays what our tables would have looked like if we only had sales data through 05JAN.

Figure 6.4 Display for Data Not Yet Available

Week Start - End: 05JAN2003 - 11JAN2003   Total Sales: $62.97   Average/Day: $62.97
* = Data Not Yet Available

Product	Sun 05JAN	Mon 06JAN	Tue 07JAN	Wed 08JAN	Thu 09JAN	Fri 10JAN	Sat 11JAN	Total
Baked potato chips	$0	*	*	*	*	*	*	$0
Barbeque potato chips	$17.88	*	*	*	*	*	*	$17.88
Classic potato chips	$31.68	*	*	*	*	*	*	$31.68
Ruffled potato chips	$13.41	*	*	*	*	*	*	$13.41
Salt and vinegar potato chips	$0	*	*	*	*	*	*	$0
WOW potato chips	$0	*	*	*	*	*	*	$0
Total	$62.97	*	*	*	*	*	*	$62.97

Week Start - End: 12JAN2003 - 18JAN2003   Total Sales: Data Not Yet Available

Product	Sun 12JAN	Mon 13JAN	Tue 14JAN	Wed 15JAN	Thu 16JAN	Fri 17JAN	Sat 18JAN	Total

Chapter 6 Summary

This chapter demonstrated how to create a weekly sales report in a calendar format, using PROC REPORT ACROSS and GROUP usage options.

- A dynamic reporting program was set up so that only start and end dates need specification for a given reporting period. The programmer can then run each weekly report without programming modifications until the next reporting period.
- The start and end dates indicate the reporting period for which the calendar data set should create records, and the period on which the sales data set should be subset.

Some other highlights of this chapter included:

- The calendar data was merged with the sales data
- PROC REPORT steps included:
    - Specifying Date as the ACROSS variable and grouping other incoming variables
    - Picture formats were used for the presentation of dates across the top of the report, and dollar amounts within the cells, including the unavailable sales data
    - A weekly (per product) Total column variable was created by a COMPUTE block
    - Total rows were created via a BREAK statement to provide each daily and the weekly sales total
    - Key summary amounts were also displayed (via a LINE statement) above each weekly table

Chapter 7: Embedding Images in a Report

Introduction 182

Example: Tables Displaying Iris Flower Measurements 182

Goals for Embedding Images in Reports 188

Source Data 188

ODS Style Templates Used 190

Programs Used 190

Implementation 190

    Setup Options, File Paths, and Image File Names 190

    Program Setup Code 191

Example 1: Obtain Images as Column of Data 192

    Code for Obtaining Images as Column of Data 193

Example 2: Repeated Images Above and Below Table 197

    Code for Repeating Images Above and Below Table 198

    Produce the Report 200

Example 3: Display Images as Column Headers 203

    Code for Displaying Images as Column Headers 204

Example 4: Display Image in Page Title 206

Code for Displaying Images in Page Titles	207
Example 5: Display Image Above Body of Table	**208**
Code for Displaying Image Above Body of Table	210
Example 6: Display Watermark on Report	**212**
Chapter 7 Summary	**213**

Introduction

The combined capabilities of ODS and PROC REPORT allow for the presentation of professional, informative reports. Reports are even further enhanced by the ability to insert images into tables. Images can increase the attractiveness of reports and improve illustrative abilities for presenting information. With the evolution of SAS and ODS, SAS programmers now have the ability to insert images in a variety of locations within a report, such as "data" within a column, before and/or after the body of a report, as column headers, and in titles and footnotes.

Example: Tables Displaying Iris Flower Measurements

Variations of a report summarizing Iris flower data are produced. The reports display measures of sepal length, sepal width, petal length, and petal width for three different types of Iris species, "Setosa," "Versicolor," and "Virginica." While the reports contain some overlapping information, the purpose of this chapter is to show how reports can be presented in different formats. Reports are created using the RTF and TAGSETS.RTF ODS destinations.

The following figures are created in this chapter:

Figure 7.1 – Displays images as a column of data.

Figure 7.2 – Displays three PROC REPORT tables in one RTF file. The table containing Iris measurements is sandwiched between two one-row tables that contain an image in every column.

Figure 7.3 – Displays images as column headers.

Figure 7.4 – Displays an image as part of a title.

Figure 7.5 – Displays an image over the table body and uses TAGSETS.RTF.

Figure 7.6 – Is the same as Figure 7.5, but with a reduced image size and a watermark added.

Figure 7.1 Images as Column of Data

Flower: Iris

SPECIES	PETAL Length	PETAL Width	SEPAL Length	SEPAL Width
Setosa	Mean: 14.62 Median: 15.00	Mean: 2.46 Median: 2.00	Mean: 50.06 Median: 50.00	Mean: 34.28 Median: 34.00
Versicolor	Mean: 42.60 Median: 43.50	Mean: 13.26 Median: 13.00	Mean: 59.36 Median: 59.00	Mean: 27.70 Median: 28.00
Virginica	Mean: 55.52 Median: 55.50	Mean: 20.26 Median: 20.00	Mean: 65.88 Median: 65.00	Mean: 29.74 Median: 30.00

Note: All measurements are reported in millimeters

Figure 7.2 Images Above and Below Table

Flower: IRIS

Measure (Mean mm)	Species		
	Setosa	Versicolor	Virginica
Petal Length	14.62	42.60	55.52
Petal Width	2.46	13.26	20.26
Sepal Length	50.06	59.36	65.88
Sepal Width	34.28	27.70	29.74

Figure 7.3 Images as Column Headers

Flower: Iris

	Setosa		Versicolor		Virginica	
Measure (mm)	Mean	Median	Mean	Median	Mean	Median
Petal Length	14.62	15.00	42.60	43.50	55.52	55.50
Petal Width	2.46	2.00	13.26	13.00	20.26	20.00
Sepal Length	50.06	50.00	59.36	59.00	65.88	65.00
Sepal Width	34.28	34.00	27.70	28.00	29.74	30.00

Figure 7.4 Image in a Page Title (appears on every page of a report)

Flower: Iris ◆ Species: Versicolor ◆ Page 1 of 2

SEPAL		PETAL	
Length (mm)	Width (mm)	Length (mm)	Width (mm)
65	28	46	15
62	22	45	15
59	32	48	18
61	30	46	14
60	27	51	16
56	25	39	11
57	28	45	13
63	33	47	16
70	32	47	14
64	32	45	15
61	28	40	13
55	24	38	11
54	30	45	15
58	26	40	12
55	26	44	12
50	23	33	10
67	31	44	14
56	30	45	15
58	27	41	10
60	29	45	15
57	26	35	10
57	29	42	13
49	24	33	10
56	27	42	13
57	30	42	12
66	29	46	13
52	27	39	14
60	34	45	16
50	20	35	10
55	24	37	10
58	27	39	12

Flower: Iris ◆ Species: Versicolor ◆ Page 2 of 2

SEPAL		PETAL	
Length (mm)	Width (mm)	Length (mm)	Width (mm)
61	29	47	14
56	29	36	13
69	31	49	15
55	25	40	13
55	23	40	13
66	30	44	14
68	28	48	14
67	30	50	17

Figure 7.5 Image Above Body of Table

FLOWER: Iris ● SPECIES: Versicolor

Versicolor Sepal and Petal Measurements			
Sepal Length (mm)	Sepal Width (mm)	Petal Length (mm)	Petal Width (mm)
65	28	46	15
62	22	45	15
59	32	48	18
61	30	46	14
60	27	51	16
56	25	39	11
57	28	45	13
63	33	47	16
70	32	47	14
64	32	45	15
61	28	40	13
55	24	38	11
54	30	45	15
58	26	40	12
55	26	44	12
50	23	33	10
67	31	44	14
56	30	45	15
58	27	41	10
60	29	45	15
57	26	35	10
57	29	42	13
49	24	33	10
56	27	42	13
57	30	42	12
66	29	46	13
52	27	39	14
60	34	45	16

(Continued)

Figure 7.6 Apply Watermark

FLOWER: Iris ❀ *SPECIES: Versicolor*

| Versicolor Sepal and Petal Measurements ||||
Sepal Length (mm)	Sepal Width (mm)	Petal Length (mm)	Petal Width (mm)
65	28	46	15
62	22	45	15
59	32	48	18
61	30	46	14
60	27	51	16
56	25	39	11
57	28	45	13
63	33	47	16
70	32	47	14
64	32	45	15
61	28	40	13
55	24	38	11
54	30	45	15
58	26	40	12
55	26	44	12
50	23	33	10
67	31	44	14
56	30	45	15
58	27	41	10
60	29	45	15
57	26	35	10
57	29	42	13
49	24	33	10
56	27	42	13
57	30	42	12
66	29	46	13
52	27	39	14
60	34	45	16
50	20	35	10
55	24	37	10
58	27	39	12
62	29	43	13
59	30	42	15
60	22	40	10

(Continued)

Goals for Embedding Images in Reports

There are a number of ways to insert images into a table when using the REPORT procedure. The coding technique differs depending on the location of the image. This chapter presents various options for embedding images within a report. The chapter also demonstrates the use of various SAS-supplied ODS Style templates.

Source Data

The source data set is the SAS supplied data set SASHELP.IRIS (Fisher's Iris Data, 1936). The data set contains information collected on the three species of IRIS named earlier: Setosa, Versicolor, and Virginica. The data collected includes species, sepal length and width, and petal length and width (all in millimeters). Table 7.1 shows a partial print of the data, and Table 7.2 displays the variable attributes of the data set.

> The iris images used in this chapter were reprinted with permission of Greg McCullough, owner of Iris City Gardens in the greater Nashville area (http://www.iriscitygardens.com).
>
> The printed book is presented in grayscale and does not display the actual colors of the images. Visit the author's web page at http://support.sas.com/publishing/authors/fine.html to see color images.

Table 7.1 Partial Print of SASHELP.IRIS Data

Species	SepalLength	SepalWidth	PetalLength	PetalWidth
Setosa	50	33	14	2
Setosa	46	34	14	3
Setosa	46	36	10	2
Setosa	51	33	17	5
Setosa	55	35	13	2
Setosa	48	31	16	2
Setosa	52	34	14	2
Setosa	49	36	14	1
Setosa	44	32	13	2
Setosa	50	35	16	6
Setosa	44	30	13	2
Setosa	47	32	16	2
Setosa	48	30	14	3
Setosa	51	38	16	2
Setosa	48	34	19	2
Setosa	50	30	16	2
Setosa	50	32	12	2
Setosa	43	30	11	1
Setosa	58	40	12	2
Setosa	51	38	19	4

Table 7.2 Contents of SASHELP.IRIS Data

#	Variable	Type	Len	Label
1	Species	Char	10	Iris Species
2	SepalLength	Num	8	Sepal Length (mm)
3	SepalWidth	Num	8	Sepal Width (mm)
4	PetalLength	Num	8	Petal Length (mm)
5	PetalWidth	Num	8	Petal Width (mm)

ODS Style Templates Used

A variety of ODS Style templates are used for the different reports.

Templates used include:

- Figure 7.1: SASWEB
- Figure 7.2: TORN
- Figure 7.3: SASWEB
- Figure 7.4: BANKER
- Figure 7.5: FANCYPRINTER
- Figure 7.6: FANCYPRINTER

Each template is specified in the ODS RTF / TAGSETS.RTF statement for a particular figure. Style overrides are applied in the TITLE and PROC REPORT statements to modify styles unique to a report.

Programs Used

All of the figures are created in one program named CH7Images.SAS.

Implementation

Setup Options, File Paths, and Image File Names

Common program features that apply to more than one figure are set up at the beginning of the program. These include:

- The ODS escape character
- The SAS system options NODATE and NONUMBER
- The output path for reports
- The image file path (input path), which is "C:\TEMP"
- Creating macro variables (for the case when species are separate variables)
- Creating formats (for the case when images correspond to the variable SPECIES' values)

Chapter 7: Embedding Images in a Report **191**

> Prior to programming, the needed images were saved to the author's "C:\TEMP\" folder.

Program Setup Code

```
** Program Setup;

ods escapechar = "^";   ❶

options nodate nonumber orientation=portrait;   ❷


** Output Path;

%let outpath = C:\Users\User\My Documents\APR\;   ❸


** Paths to Images as Macro Variables;   ❹

%let setosa      = "c:\temp\iSetosacp10.png";

%let versicolor  = "c:\temp\iVersicolorcp10.png";

%let virginica   = "c:\temp\iVirginicacp10.png";

%let versicolorw = "c:\temp\iVersicolorsm30.jpg";

%let versicolorw2 = "c:\temp\iVersicolorsm20.jpg";


** Paths to Images as Formats;   ❺

proc format;
  value $flower
    "Setosa"     = "c:\temp\iSetosacp10.png"
    "Versicolor" = "c:\temp\iVersicolorcp10.png"
    "Virginica"  = "c:\temp\iVirginicacp10.png";
run;
```

❶ We need to declare the ODS escape character before we can use it for the ODS functions used throughout the chapter code. The **ODS ESCAPECHAR is declared as the caret symbol ("^")**.

❷ NODATE suppresses the default date printed above SAS output. NONUMBER suppresses the SAS page numbers. The page orientation is set as portrait.

❸ The output path to which all reports will be sent is specified in the macro variable OUTPATH.

❹ Macro variables are created so we do not have to repeatedly type the file paths and names.

❺ Likewise, the $FLOWER format is created so that we do not have to repeatedly type the file paths and names.

Example 1: Obtain Images as Column of Data

Figure 7.1 Images as Column of Data

Flower: Iris

SPECIES	PETAL Length	PETAL Width	SEPAL Length	SEPAL Width
Setosa	Mean: 14.62 Median: 15.00	Mean: 2.46 Median: 2.00	Mean: 50.06 Median: 50.00	Mean: 34.28 Median: 34.00
Versicolor	Mean: 42.60 Median: 43.50	Mean: 13.26 Median: 13.00	Mean: 59.36 Median: 59.00	Mean: 27.70 Median: 28.00
Virginica	Mean: 55.52 Median: 55.50	Mean: 20.26 Median: 20.00	Mean: 65.88 Median: 65.00	Mean: 29.74 Median: 30.00

Note: All measurements are reported in millimeters

The first example, Figure 7.1, portrays images in the Species column of the report. The key tasks include:

- Creating the $FLOWER format (which has already been done in the Program Setup Code).
- Applying the $FLOWER format to the PREIMAGE= attribute in the SPECIES DEFINE statement.

Code for Obtaining Images as Column of Data

```
** IMAGES AS COLUMN DATA;  ❶

ods _all_ close;

ods rtf style=sasweb file="&outpath.Ch7_column.rtf"
bodytitle;


title height=16 pt color=CX6D5299 bold font=Garamond "Flower: Iris";  ❷


footnote color=black bold justify=left font=Garamond height=11 pt
        "^S={asis=on}                Note: All measurements are
reported in millimeters";  ❸


proc report data=sashelp.iris nowd missing split="|" center
   style(header)=[background=CX6D5299 font_size=11 pt font_weight=bold
              font_face=Garamond]
   style(column)=[cellwidth=1.1 in vjust=m just=d font_weight=bold
              font_face=Garamond font_size=11 pt];  ❹


** Create Aliases so that Multiple Statistics can be Reported;
column ("SPECIES" species)


       petallength petallength=plength2
       petalwidth  petalwidth=pwidth2
       ("PETAL" ("Length" plmeanmed) ("Width" pwmeanmed))


       sepallength sepallength=slength2
       sepalwidth  sepalwidth=swidth2
       ("SEPAL" ("Length" slmeanmed) ("Width" swmeanmed));  ❺
```

```
   define species      / " " group style(column)=[preimage=$flower.
                         protectspecialchars=off cellwidth=1.24 in
vjust=m];  ❻
```

```
** Statistics;  ❼
define petallength  /  mean noprint;
define plength2     /  median noprint;
define petalwidth   /  mean noprint;
define pwidth2      /  median noprint;
define sepallength  /  mean noprint;
define slength2     /  median noprint;
define sepalwidth   /  mean noprint;
define swidth2      /  median noprint;


** Print Computed Variables;
define plmeanmed    / "" computed ;
define pwmeanmed    / "" computed ;
define slmeanmed    / "" computed ;
define swmeanmed    / "" computed ;


** COMPUTED (And Printed) Columns;  ❽
compute plmeanmed /char length=60;
   plmeanmed= "Mean: " || strip(put(petallength.mean,8.2)) ||"^n" ||
              "Median: " || strip(put(plength2,10.2));
endcomp;


compute pwmeanmed /char length=60;
   pwmeanmed= "Mean: " || strip(put(petalwidth.mean,8.2)) ||"^n" ||
```

```
                "Median: " || strip(put(pwidth2,10.2));
   endcomp;


   compute slmeanmed /char length=60;
      slmeanmed= "Mean: " || strip(put(sepallength.mean,8.2)) ||"^n" ||
                "Median: " || strip(put(slength2,10.2));
   endcomp;


   compute swmeanmed /char length=60;
      swmeanmed= "Mean: " || strip(put(sepalwidth.mean,8.2)) ||"^n" ||
                "Median: " || strip(put(swidth2,10.2));
   endcomp;
run;
ods _all_ close;
ods html;
title;
footnote;
```

❶ SASWEB is chosen as the ODS Style Template to arrive at the overall table look. The BODYTITLE option is added to the ODS RTF statement so that titles and footnotes will appear directly above and below the table, rather than in the header and footer sections of the page.

❷ The style of the title is modified by applying the height, color, and font options. The HEIGHT= option specifies the point size for the title, in this case, 16 pt. The font color is specified as the RGB color code CX6D5299, which represents "Light purplish blue." An RGB color code list is provided in http://support.sas.com/techsup/technote/ts688/ts688.html. The font weight and type are specified as bold Garamond.

❸ The spaces prior to the text in the footnote are intentional, so the text will line up with the table. The ASIS=ON style attribute requested with the style function is necessary to preserve the leading spaces.

❹ Style overrides for headers and columns are defined in the PROC REPORT statement. The header background is set to the color of "Light purplish blue" (CX6D5299). The font face is changed to Garamond and bolded. The font size is set to 11 pt.

The column data has additional specifications. The vertical justification of data is specified as shown in the middle of the cell (with vjust=m). The width of the cells is specified as 1.1 inches. The horizontal justification is decimal aligned (just=d). The font face is Garamond and the font weight is set to bold.

❺ Aliases are created for each of the four original measurement variables (petallength, petalwidth, sepallength, sepalwidth) so multiple statistics (i.e. mean and median) can be produced for each variable. Spanning headers are created around the variables to be printed.

❻ The PREIMAGE= attribute is added to the Species DEFINE statement column(style) specification to add the iris images. The format $FLOWER is added to apply the appropriate image to each species value (Setosa, Versicolor, and Virginica).

❼ We use a set of NOPRINTED variables to obtain statistics:

- Means and medians are obtained for the petal and sepal lengths and widths by specifying either MEAN or MEDIAN as the statistic for these ANALYSIS variables.

- Means use the original variable name (i.e. PETALLENGTH, PETALWIDTH, SEPALLENGTH, SEPALWIDTH) and medians use the alias name (i.e. PLENGTH2, PWIDTH2, SLENGTH2, SWIDTH2) declared in the COLUMN statement. The use of the alias names allows us to calculate more than one statistic (in this case, median) for the same variable.

- We suppress the printing of the individual mean and median columns because we will print COMPUTE block variables which create custom character strings containing the needed information. The COMPUTED variables stack the mean and median within each petal and sepal cell, as shown in the following example for the Setosa Species.

	PETAL		SEPAL	
SPECIES	Length	Width	Length	Width
Setosa	Mean: 14.62 Median: 15.00	Mean: 2.46 Median: 2.00	Mean: 50.06 Median: 50.00	Mean: 34.28 Median: 34.00

❽ The custom character strings are created via COMPUTE blocks. This process is needed for Petal Length, Petal Width, Sepal Length, and Sepal Width (thus four COMPUTE blocks). Each character string concatenates the text "Mean", a character version of the mean value, the newline function (specified by our escape character followed by "n"), the word "Median," and the character version of the median value. These computed variables are printed in the final report.

Example 2: Repeated Images Above and Below Table

This section describes how to create Figure 7.2, which treats images as variables (IMG1, IMG2, and IMG3) and displays them as repeated columns in the REPORT procedure above and below the measurements table.

Figure 7.2 Images Above and Below Table

Flower: IRIS

Measure (Mean mm)	Species		
	Setosa	Versicolor	Virginica
Petal Length	14.62	42.60	55.52
Petal Width	2.46	13.26	20.26
Sepal Length	50.06	59.36	65.88
Sepal Width	34.28	27.70	29.74

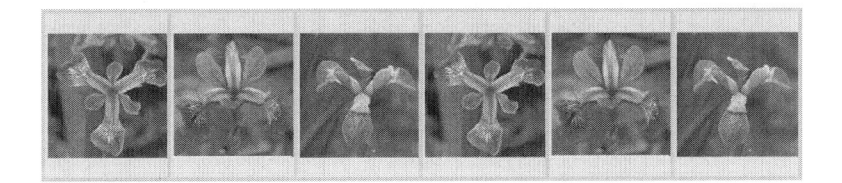

Highlights of this section include:

- Creating a one observation data set that contains the three images as variables (named IMG1, IMG2, and IMG3).
- Transposing the IRIS data set to contain two variables (named PARAM and MEAS) and a record for each measurement (Petal Length, Petal Width, Sepal Length, and Sepal Width).

- Running three PROC REPORTS sandwiched in the same ODS RTF report
  - PROC REPORT, repeating columns of IMG1, IMG2, IMG3
  - PROC REPORT of the transposed IRIS data
  - PROC REPORT, repeating columns of IMG1, IMG2, IMG3 (repeat of the first PROC REPORT).

Code for Repeating Images Above and Below Table

```
** Create Data Set with One Observation and Three Images; ❶
** Single quotes used on outside since macro variable paths use double
   quotes;
data img;
length img1-img3 $100. ;
   img1='^S={postimage=&setosa}';
   img2='^S={postimage=&versicolor}';
   img3='^S={postimage=&virginica}';
   blank=" ";
run;




** Create Transposed Iris Data Set with Parameter Identifier; ❷
data tiris(drop=sepal: petal:);
  length descrip $16;
  format meas 8.2;

  set sashelp.iris;

  descrip = "Petal Length"; meas = petallength; output;
  descrip = "Petal Width";  meas = petalwidth;  output;
```

```
      descrip = "Sepal Length"; meas = sepallength; output;
      descrip = "Sepal Width";  meas = sepalwidth;  output;
   run;
```

❶ The one observation data set named IMG is created. Because the variables IMG1, IMG2, and IMG3 contain the escape character and inline style function (S={}), these variables will be rendered as images in the ODS destination. Note that single quotes are used to surround the style function because double quotes were used in the macro variables defined in the earlier %LET statements. As an example, what we're really setting IMG1 to is

```
img1='^S={postimage="c:\temp\iSetosacp10.png"}';
```

❷ The IRIS data set is transposed so that we can get each measurement along with a description as its own record. Table 7.3 displays a partial print of the new data set named TIRIS.

Table 7.3 Partial PRINT of Transposed IRIS

descrip	Species	Meas
Sepal Length	Setosa	50
Sepal Width	Setosa	33
Petal Length	Setosa	14
Petal Width	Setosa	2
Sepal Length	Setosa	46
Sepal Width	Setosa	34
Petal Length	Setosa	14
Petal Width	Setosa	3
Sepal Length	Setosa	46
Sepal Width	Setosa	36
Petal Length	Setosa	10
Petal Width	Setosa	2
Sepal Length	Setosa	51
Sepal Width	Setosa	33
Petal Length	Setosa	17

200 PROC REPORT by Example: Techniques for Building Professional Reports Using SAS

descrip	Species	Meas
Petal Width	Setosa	5
Sepal Length	Setosa	55
Sepal Width	Setosa	35
Petal Length	Setosa	13
Petal Width	Setosa	2
Sepal Length	Setosa	48

Produce the Report

```
** REPEATED IMAGES ABOVE AND BELOW TABLE;
ods _all_ close;
ods rtf style=torn file="&outpath.Ch7sndwch.rtf" startpage=no;  ❶


title "Flower: IRIS";


** FIRST PROC REPORT OF IMG DATA SET;  ❷
proc report data=img nowd missing split="|" center
    style=[frame=void rules=none protectspecialchars=off]
    style(column)=[font_size=8 pt just=c cellwidth=1 in];

    column img1 img2 img3 img1 img2 img3;
    define img1 / "";
    define img2 / "";
    define img3 / "";
run;
```

Chapter 7: Embedding Images in a Report **201**

```
** PROC REPORT FOR DATA TABLE; ❸
proc report data=tiris nowd missing split="|" center
   style=[frame=void]
   style(header)=[font_size=14 pt font_weight=bold background=white]
   style(column)=[cellwidth=1.3 in font_size=10 pt just=c]
   out=PROUT;


column   ("Measure|(Mean mm)" descrip)
         ("Species" species), (meas meas=meas2);


   define species / across "";
   define descrip / group   "" style(column header)=[just=l cellwidth=2.1 in];
   define meas       /  mean     "";
   define meas2      /  median noprint;   /** used for Figure 7.3 **/
run;
```

```
** SECOND PROC REPORT OF IMG DATA SET; ❹
proc report data=img nowd missing split="|" center
   style=[frame=void rules=none protectspecialchars=off]
   style(column)=[font_size=8 pt just=c cellwidth=1 in];


   column img1 img2 img3 img1 img2 img3;
   define img1 / "";
   define img2 / "";
   define img3 / "";
run;
ods _all_ close;
```

```
ods html;
title;
footnote;
```

❶ For Figure 7.2, TORN is declared as the ODS Style Template to use. The STARTPAGE option is set to "no" because without this setting, the three separate REPORT procedures would each appear on their own page.

❷ The first PROC REPORT repeatedly displays the one record data set IMG variables as columns with the statement

```
COLUMN img1 img2 img3 img1 img2 img3;
```

❸ The second PROC REPORT provides the data table. The DESCRIP column is DEFINEd as GROUP, and the SPECIES column as ACROSS, so that Species values (Setosa, Versicolor, Virginica) become columns that contain Petal Length, Petal Width, Sepal Length, and Sepal Width measurements.

❹ The third PROC REPORT again repeatedly displays the one record data set IMG variables as columns with the statement

```
COLUMN img1 img2 img3 img1 img2 img3;
```

Note that an output data set (PROUT) is created and medians are derived (though not printed in Figure 7.2). The PROUT data set is the source data set for the next figure, Figure 7.3.

Example 3: Display Images as Column Headers

This section demonstrates how to create Figure 7.3, which shows images as column headers.

Figure 7.3 Images as Column Headers

Measure (mm)	Setosa Mean	Setosa Median	Versicolor Mean	Versicolor Median	Virginica Mean	Virginica Median
Petal Length	14.62	15.00	42.60	43.50	55.52	55.50
Petal Width	2.46	2.00	13.26	13.00	20.26	20.00
Sepal Length	50.06	50.00	59.36	59.00	65.88	65.00
Sepal Width	34.28	34.00	27.70	28.00	29.74	30.00

Note that the input for the REPORT procedure is the PROC REPORT output dataset created in the boxed code from Example 2. Table 7.4 shows a print of the data set PROUT.

Table 7.4 PROC REPORT OUTPUT DATA SET PROUT

Descrip	_C2_	_C3_	_C4_	_C5_	_C6_	_C7_	_BREAK_
Petal Length	14.62	15.00	42.60	43.50	55.52	55.50	
Petal Width	2.46	2.00	13.26	13.00	20.26	20.00	
Sepal Length	50.06	50.00	59.36	59.00	65.88	65.00	
Sepal Width	34.28	34.00	27.70	28.00	29.74	30.00	

As shown in Table 7.4, the ACROSS columns in the output data set are in the form _Cn_, where n indicates the column number. In this case, columns _C2_ and _C3_ (Column 2 and Column 3) represent Setosa Mean and Setosa Median, respectively. _C4_ and _C5_ (Column 4 and Column 5) represent Versicolor Mean and Median, respectively. _C6_ and _C7_ (Column 6 and Column 7) represent Virginica Mean and Median, respectively.

Code for Displaying Images as Column Headers

```
** IMAGES AS COLUMN HEADERS;

ods _all_ close;

ods rtf style=sasweb file="&outpath.Ch7_header.rtf"; ❶

title height=20 pt bold italic bcolor=CX6D5299 color=white
font=Georgia "Flower: Iris"; ❷

proc report data=PROUT nowd missing split="|"
   style=[protectspecialchars=off cellspacing=5]
   style(header)=[font_weight=bold font_size=12 pt background=white
                  foreground=CX483D8B]
   style(column) = [cellwidth=.8 in just=c font_size=10 pt]; ❸

   ** Insert Images as Spanning Headers;
   column ("Measure (mm)" descrip)

     ('^S={pretext="Setosa" postimage=&setosa}'

       ("Mean" _c2_) ("Median" _c3_)
     )


     ('^S={pretext="Versicolor" postimage=&versicolor}'

       ("Mean" _c4_) ("Median" _c5_)
     )


     ('^S={pretext="Virginica" postimage=&virginica}'

       ("Mean" _c6_)("Median" _c7_ )
     ); ❹

   ** DEFINE Specifications; ❺
   define descrip / order ""
```

```
                         style(column)=[just=l cellwidth=1.6 in indent=.2
in];
   define _c2_ / "";
   define _c3_ / "";
   define _c4_ / "";
   define _c5_ / "";
   define _c6_ / "";
   define _c7_ / "";
run;
ods _all_ close;
ods html;
title;
footnote;
```

❶ Figure 7.3 uses the SASWEB ODS Style Template.

❷ The title style is modified by applying the height (font size), bold, italic, bcolor (background color) color (foreground color) and font (font type) options. The background color uses RGB code CX6D5299, which is "Light purplish blue."

❸ Style overrides for the overall output, headers and columns are specified in the PROC REPORT statement. For the overall report, PROTECTSPECIALCHARS=OFF is specified so that SAS and ODS do not try to "protect" the backslash characters found in the images paths. The cell spacing is increased to provide thicker borders between the cells.

Header font is set to bold 12 pt. font. The background color is changed from the default SASWEB blue to white. The font color is changed from the default SASWEB white to "DarkSlateBlue" (foreground=CX483D8B).

Column cellwidth is set to .8 inches, column data is centered, and its font is changed to 10 pt. (from the default SASWEB 9.5 pt.).

❹ The images are inserted as Spanning Headers in the COLUMN statement. The PRETEXT= and POSTIMAGE= options are used within ODS style functions. The PRETEXT= option is used to add the species name prior to the species image (POSTIMAGE). The species macro variables are called for the POSTIMAGEs, for example, postimage= &setosa resolves to postimage= "c:\temp\iSetosacp10.png")

206 PROC REPORT by Example: Techniques for Building Professional Reports Using SAS

❺ The DESCRIP column style is formatted slightly different than the other columns. The data is left justified, given a larger cell width, and indented .2 inches from the left. All column labels are set to null in the DEFINE statements so the variable names are not printed.

Example 4: Display Image in Page Title

This section demonstrates how to create Figure 7.4, which displays an image in the page title. Because the image is part of the page title, the image appears on every page of the report. The PREIMAGE= attribute is used to obtain an image before the title text. Although not shown, the same technique can be used for page footnotes.

Figure 7.4 Image in a Page Title (partial print of Pages 1 and 2)

Flower: Iris ◆ Species: Versicolor ◆ Page 1 of 2

SEPAL		PETAL	
Length (mm)	Width (mm)	Length (mm)	Width (mm)
65	28	46	15
62	22	45	15
59	32	48	18
61	30	46	14
60	27	51	16
56	25	39	11
57	28	45	13
63	33	47	16
70	32	47	14
64	32	45	15
61	28	40	13
55	24	38	11
54	30	45	15
58	26	40	12
55	26	44	12
50	23	33	10
67	31	44	14
56	30	45	15
58	27	41	10
60	29	45	15
57	26	35	10
57	29	42	13
49	24	33	10
56	27	42	13
57	30	42	12
66	29	46	13
52	27	39	14
60	34	45	16
50	20	35	10
55	24	37	10
58	27	39	12

Flower: Iris ◆ Species: Versicolor ◆ Page 2 of 2

SEPAL		PETAL	
Length (mm)	Width (mm)	Length (mm)	Width (mm)
61	29	47	14
56	29	36	13
69	31	49	15
55	25	40	13
55	23	40	13
66	30	44	14
68	28	48	14
67	30	50	17

Code for Displaying Images in Page Titles

```
** IMAGE IN TITLE;
ods _all_ close;
ods rtf file="&outpath.Ch7_titleimg.rtf" style=banker;   ❶


title h=14 pt '^S={preimage=&versicolor}
  Flower: Iris ^{style [font_face=wingdings]u}
  Species: Versicolor ^{style [font_face=wingdings]u}
  Page ^{thispage} of ^{lastpage}';   ❷


proc report data=sashelp.iris(where=(species="Versicolor")) nowd
missing
  split="|" center
  style(header)=[just=center]
  style(column)=[just=center cellwidth=1.2 in];


  column species ("SEPAL" sepallength sepalwidth)
                 ("PETAL" petallength petalwidth);


  define species / "Species" NOPRINT ;
```

208 PROC REPORT by Example: Techniques for Building Professional Reports Using SAS

```
        define sepallength / "Length (mm)"    format=8.;
        define sepalwidth  / "Width  (mm)"    format=8.;
        define petallength / "Length (mm)"    format=8.;
        define petalwidth  / "Width  (mm)"    format=8.;
run;
ods _all_ close;
ods html;
title;
footnote;
```

❶ BANKER is the ODS Style Template for this example.

❷ A style function places the PREIMAGE= option directly in the TITLE statement. This style is started and ended with brackets ("{}").

Additional style functions are nested within the same TITLE statement. The ^{style [font_face=wingdings]u} creates the diamond symbol, and this is placed between the "Flower: Iris", "Species: Versicolor" and "Page" text.

The "Page" text is followed by two other ODS functions: {thispage}, the text "of", and the function {lastpage}.

Example 5: Display Image Above Body of Table

This example demonstrates how to create Figure 7.5, which shows how to insert an image above the body of the PROC REPORT table. Unlike Example 4, for which the image prints on every page, this example prints the image only once, before the table. The image is inserted with the STYLE= attribute (rather than a STYLE function) within the PROC REPORT statement. The PREIMAGE= attribute is nested within the STYLE= attribute. Although not shown, the POSTIMAGE= attribute can be used to insert the image below the table body. The "(Continued)" text at the bottom of Figure 7.5 is part of the report and is obtained with ODS TAGSETS.RTF.

Figure 7.5 Image Above Body of Table

FLOWER: Iris ● *SPECIES: Versicolor*

Versicolor Sepal and Petal Measurements			
Sepal Length (mm)	Sepal Width (mm)	Petal Length (mm)	Petal Width (mm)
65	28	46	15
62	22	45	15
59	32	48	18
61	30	46	14
60	27	51	16
56	25	39	11
57	28	45	13
63	33	47	16
70	32	47	14
64	32	45	15
61	28	40	13
55	24	38	11
54	30	45	15
58	26	40	12
55	26	44	12
50	23	33	10
67	31	44	14
56	30	45	15
58	27	41	10
60	29	45	15
57	26	35	10
57	29	42	13
49	24	33	10
56	27	42	13
57	30	42	12
66	29	46	13
52	27	39	14
60	34	45	16

(Continued)

Code for Displaying Image Above Body of Table

```
** IMAGE ABOVE BODY OF REPORT;

ods _all_ close;

ods tagsets.rtf file="&outpath.Ch7_bodyimg.rtf" style=fancyprinter;   ❶


title h=18 pt " FLOWER: Iris   ^{style [font_face=wingdings
color=darkpurpleblue]|}   SPECIES: Versicolor";   ❷


proc report data=sashelp.iris(where=(species="Versicolor")) nowd
missing
    split="|" center
    style(report)=[preimage=&versicolorw cellpadding=0 cellspacing=0]
    style(column)=[just=center cellwidth=1.15 in font_size=11 pt]
    style(header)=[font_size=11 pt font_weight=bold];   ❸


  column
    species
    ("^{newline}^{style [font_weight=bold]
        Versicolor Sepal and Petal Measurements}^{newline}"
      sepallength sepalwidth petallength petalwidth
    );   ❹


  define species / noprint;   ❺
run;
ods _all_ close;
ods html;
title;
footnote;
```

❶ Note that TAGSETS.RTF is the ODS destination specified. This is because we want the default TAGSETS.RTF "Continued" note to appear at the bottom of each page until the last page of the report.

The ODS Style Template "FancyPrinter" provides the template for this report.

❷ A flower symbol is inserted between the Flower and Species descriptions by applying wingdings font face to the pipe ("|") character with a style function.

❸ The PREIMAGE= attribute is placed in a style(report)= attribute as part of the PROC REPORT statement.

❹ Though SPECIES is not printed (because this report was subset on only one species, "Versicolor") this character variable is included so the four ANALYSIS variables (sepallength, sepalwidth, petallength, and petalwidth) are not summed. (The addition of the DISPLAY variable prevents the ANALYSIS variables from being summed. By default, the character variable SPECIES is DISPLAY, and the numeric variables SEPALLENGTH, SEPALWIDTH, PETALLENGTH, and PETALWIDTH are ANALYSIS).

❺ Suppress the printing of the SPECIES column with NOPRINT.

Example 6: Display Watermark on Report

Another feature available with ODS TAGSETS.RTF is a watermark option, as shown in Figure 7.6.

Figure 7.6 Apply Watermark

FLOWER: Iris ● *SPECIES: Versicolor*

Versicolor Sepal and Petal Measurements			
Sepal Length (mm)	Sepal Width (mm)	Petal Length (mm)	Petal Width (mm)
65	28	46	15
62	22	45	15
59	32	48	18
61	30	46	14
60	27	51	16
56	25	39	11
57	28	45	13
63	33	47	16
70	32	47	14
64	32	45	15
61	28	40	13
55	24	38	11
54	30	45	15
58	26	40	12
55	26	44	12
50	23	33	10
67	31	44	14
56	30	45	15
58	27	41	10
60	29	45	15
57	26	35	10
57	29	42	13
49	24	33	10
56	27	42	13
57	30	42	12
66	29	46	13
52	27	39	14
60	34	45	16
50	20	35	10
55	24	37	10
58	27	39	12
62	29	43	13
59	30	42	15
60	22	40	10

(Continued)

The Example 6 program requires three changes to the Example 5 program:

- Change the output file name so that it does not overwrite Example 5.

- Include the WATERMARK= option in the ODS TAGSETS.RTF statement:

  ```
  ods tagsets.rtf file="c:\temp\Ch7_wmbodyimg.rtf"
  style=fancyprinter
      options (watermark="DRAFT");
  ```

  While the word "DRAFT" is chosen here, the programmer can choose the text to serve as the watermark.

- Use a reduced size image so that the watermark covers the entire table. We use the image stored in the macro variable VERSICOLORW2 in the PROC REPORT statement.

  ```
  style(report)=[preimage=&versicolorw2 cellpadding=0
  cellspacing=0]
  ```

Chapter 7 Summary

Table 7.5 summarizes some of the tasks that were performed to embed images in this chapter's report examples.

Table 7.5 Summary of Tasks

Task	Summary
Setup Image Paths and File Names	The needed images were already saved to the author's folder ("C:\TEMP").
	The ODS Escapechar was declared.
	Options to be used for all reports were specified.
	Needed macro variables and formats were created.
Display Images in Column Data	A format was created to attach each image to a flower species name. The format was applied to the Species column in the SPECIES DEFINE statement.
Repeated Images Above and Below Table	A one observation data set containing three variables was created. Assignment statements set each variable to its corresponding image along with the needed style function specifications, for example img1='^S={postimage="c:\temp\iSetosacp10.png"}';

Task	Summary
	PROC REPORT was invoked three times, with the IRIS measurements data sandwiched between the images ("IMG") data set.
Display Images as Column Headers	PROC REPORT was used to transpose the IRIS data and produce a data set in which the species are columns in the form _C2_, _C3_, and _C4_ ... _C7. Using style functions with the PREIMAGE= attribute, the images were specified as spanning headers in the COLUMN statement, for example ('^S={pretext="Setosa" postimage=&setosa}' ("Mean" _C2_) ("Median" _C3_))
Display Image in Page Title	Using a style function again, the images were placed directly in the title statement. A simplified example is the title '^S={preimage="c:\temp\iVersicolorcp10.png"} Flower: Iris';
Display Image Above the Table Body	The image is specified in the PROC REPORT statement, for example, proc report data=sashelp.iris nowd missing split='|' center style=[preimage="c:\temp\iVersicolorsm30.jpg"];
A Watermark is Possible With ODS TAGSETS.RTF	For example, ods tagsets.rtf file="c:\temp\bodywmark.rtf" style=fancyprinter options (watermark="DRAFT");

Chapter 8: Combining Graphs and Tabular Data

Introduction 216

Example: Dashboard Report of Shoe Sales... 216

Goals for Creating the Shoe Sales Dashboard 218
    Key Steps 218

Source Data 218

ODS Style Template Used 219

Programs Used... 219

Implementation 220

Create a Summary Data Set using PROC REPORT 220
    Code for Creating a Summary Data Set 220

Obtain Regional Ranking Information.. 222
    Code for Obtaining Regional Ranking Information.. 222

Create a New ODS Style Template 223

Create the ODS LAYOUT for the Report 226

Create Formats Needed for Outputs 226

Use PROC SGPLOT to Create Vertical Bar Charts 227

| Code for SGPLOT Vertical Bar Charts | 227 |

Using PROC SGPLOT to Create a Horizontal Bar Chart 230

| Horizontal Bar Chart Code | 230 |

Using PROC REPORT to Obtain Tabular Output 231

Using PROC SGPANEL to Create Bar Charts for the Top 3 Regions 232

Chapter 8 Summary 235

Introduction

ODS LAYOUT provides the ability to customize reports as never before with SAS. We can present tables, graphics, images, and text on the same page. We can arrange the location of each piece of output on the page. In sum, we have much more control over the reports we develop.

It should be noted that ODS Layout is a preproduction feature of SAS versions 9.0 through 9.3, and will be in full production in SAS version 9.4. This book is written using SAS 9.3 and the chapter focuses on ODS LAYOUT for a PDF file.

Example: Dashboard Report of Shoe Sales

A PDF report is produced to display Shoe Sales, Returns, Percentage of Sales by Region, and Sales by Product for the "Top Three" regions. Three graphs and one table are presented in a Dashboard report format.

Figure 8.1 displays the PDF report.

Chapter 8: Combining Graphs and Tabular Data   217

Figure 8.1 Shoe Sales Dashboard Report

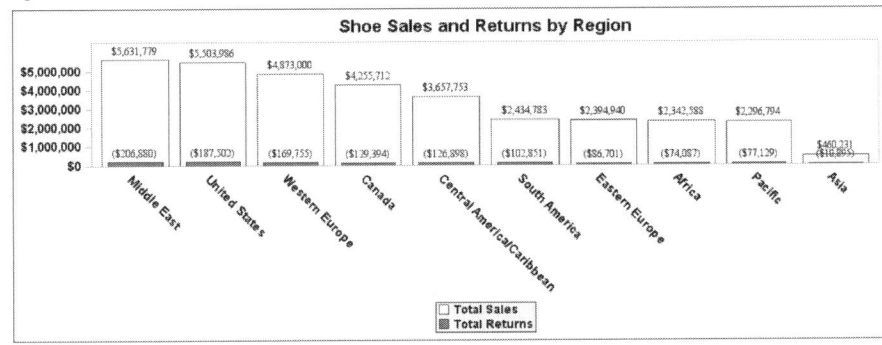

Shoe Sales and Returns by Region

Region	Total Sales	Total Returns
Middle East	$5,631,779	($206,880)
United States	$5,503,986	($187,502)
Western Europe	$4,873,000	($169,755)
Canada	$4,255,712	($139,394)
Central America/Caribbean	$3,657,753	($126,898)
South America	$2,434,783	($102,851)
Eastern Europe	$2,394,940	($86,701)
Africa	$2,342,588	($74,087)
Pacific	$2,296,794	($77,129)
Asia	$460,231	($10,895)

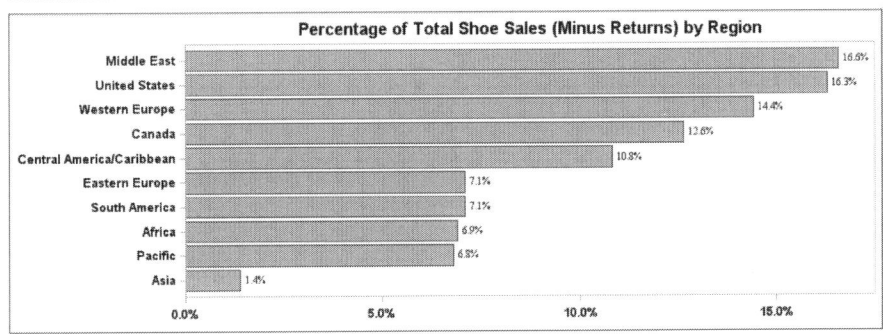

Percentage of Total Shoe Sales (Minus Returns) by Region

- Middle East: 16.6%
- United States: 16.3%
- Western Europe: 14.4%
- Canada: 12.6%
- Central America/Caribbean: 10.8%
- Eastern Europe: 7.1%
- South America: 7.1%
- Africa: 6.9%
- Pacific: 6.8%
- Asia: 1.4%

Region	Sales	Returns	Sales Minus Returns	Percentage of Total
Middle East	$5,631,779	$206,880	$5,424,899	16.6%
United States	$5,503,986	$187,502	$5,316,484	16.3%
Western Europe	$4,873,000	$169,755	$4,703,245	14.4%
Canada	$4,255,712	$129,394	$4,126,318	12.6%
Central America/Caribbean	$3,657,753	$126,898	$3,530,855	10.8%
South America	$2,434,783	$102,851	$2,331,932	7.1%
Eastern Europe	$2,394,940	$86,701	$2,308,239	7.1%
Africa	$2,342,588	$74,087	$2,268,501	6.9%
Pacific	$2,296,794	$77,129	$2,219,665	6.8%
Asia	$460,231	$10,895	$449,336	1.4%

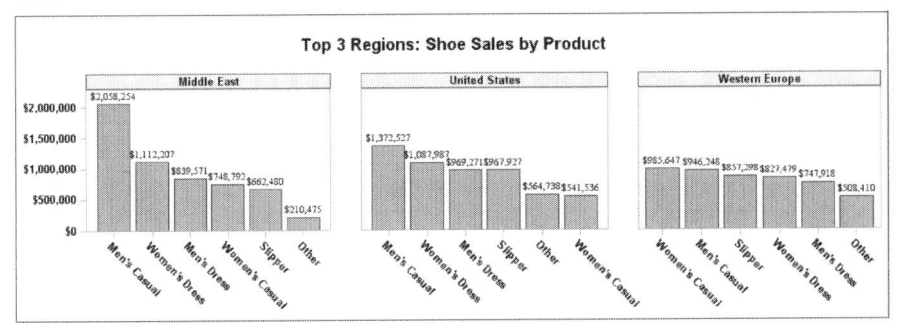

Top 3 Regions: Shoe Sales by Product

Middle East:
- Men's Casual: $2,058,254
- Women's Dress: $1,112,207
- Men's Dress: $839,571
- Women's Casual: $748,792
- Slipper: $662,480
- Other: $210,475

United States:
- Men's Casual: $1,322,527
- Women's Dress: $1,087,987
- Men's Dress: $969,271
- Women's Casual: $967,927
- Slipper: $564,738
- Other: $541,536

Western Europe:
- Women's Casual: $985,647
- Men's Casual: $946,248
- Men's Dress: $857,298
- Slipper: $827,439
- Women's Dress: $747,918
- Other: $508,410

Goals for Creating the Shoe Sales Dashboard

The goals for creating the Dashboard report include designing the layout, making necessary data set modifications, and running a series of procedures whose outputs will be placed into the layout.

Key Steps

Key steps include:

- Declare ODS LAYOUT START and END for a set of graphs and tables to be placed on the page.
  - This report uses COLUMNS= and ROWS= options to define how the page will be gridded.
- PROC REPORT is used to create a source data set for two of the graphs and the table.
- PROC RANK is used to easily add a ranking variable needed for the final graph.
- SG plots (PROC SGPLOT and PROC SGPANEL) are used to produce the graphs.
- PROC REPORT is used again to produce a printed table.

Source Data

The source data set is the SASHELP.SHOES data set. Several variables are kept for the purpose of this example. These include Region, Product, Sales, and Returns. Table 8.1 displays a partial print, and Table 8.2 displays partial contents of the SASHELP.SHOES data set.

Table 8.1 Partial Print of SASHELP.SHOES Data

Region	Product	Sales	Returns
Africa	Boot	$29,761	$769
Africa	Men's Casual	$67,242	$2,284
Africa	Men's Dress	$76,793	$2,433
Africa	Sandal	$62,819	$1,861
Africa	Slipper	$68,641	$1,771
Africa	Sport Shoe	$1,690	$79
Africa	Women's Casual	$51,541	$940
Africa	Women's Dress	$108,942	$3,233
Africa	Boot	$21,297	$710
Africa	Men's Casual	$63,206	$2,221
Africa	Men's Dress	$123,743	$3,621

Region	Product	Sales	Returns
Africa	Sandal	$29,198	$1,530
Africa	Slipper	$64,891	$1,823
Africa	Sport Shoe	$2,617	$168
Africa	Women's Dress	$90,648	$2,690
Africa	Boot	$4,846	$229
Africa	Men's Casual	$360,209	$9,424
Africa	Men's Dress	$4,051	$97
Africa	Sandal	$10,532	$598
Africa	Slipper	$13,732	$1,216

Table 8.2 Partial Contents of SASHELP.SHOES Data

#	Variable	Type	Len	Format	Informat	Label
1	Region	Char	25			
2	Product	Char	14			
3	Sales	Num	8	DOLLAR12.	DOLLAR12.	Total Sales
4	Returns	Num	8	DOLLAR12.	DOLLAR12.	Total Returns

ODS Style Template Used

The ODS Style Template MEADOW is modified and saved to a new template named MEADOWG.

Programs Used

The program is named Ch8Graph.sas.

Implementation

Create a Summary Data Set using PROC REPORT

A new data set named SALES is created to summarize total sales, total returns, sales minus returns, and percent of total sales by REGION. PROC REPORT is used to create this data set. The SALES data will be the source of the first three outputs (two graphs and one table) in the ODS LAYOUT.

Code for Creating a Summary Data Set

```
** Create new data set via PROC REPORT to be fed into graphs and
table;
proc report data=sashelp.shoes(keep=region sales returns) nowd missing
   OUT = SALES(where=(_BREAK_=" "));  ❶

   column region sales returns finamt cumfin pct;  ❷


   define region /group "Region";  ❸


   ** Get Cumulative Total;

   rbreak before   /summarize;  ❹

   ** Create COMPUTEd Variables;

   ** Sales - Returns;

   compute finamt;  ❺
      if nmiss(returns.sum,sales.sum)=0 then
         finamt=sales.sum-returns.sum;
   endcomp;

   ** Grand Total Sales - Returns;

   compute cumfin;  ❻
```

```
        if _break_ = "_RBREAK_" then dstot=finamt;
     cumfin=dstot;
  endcomp;

  ** Region Percent of Grand Total;
  compute pct;  ❼
     pct= round((finamt/cumfin) * 100,.1);
  endcomp;
run;
```

❶ PROC REPORT is run on the SASHELP.SHOES data set. The output data set is named SALES. Only detail rows are kept (i.e. where _BREAK_=" ").

❷ The last three variables in the COLUMN statement (FINAMT, CUMFIN, and PCT) are computed variables.

❸ REGION is grouped so that rows represent REGION summaries.

❹ The RBREAK BEFORE statement is used to obtain totals across all regions. The Total Sales – Total Returns (i.e. where _break_ = _RBREAK_) amount is copied to the variable CUMFIN. This amount is used as the denominator for the derivation of PCT, Percent of (Total Sales – Total Returns).

❺ FINAMT represents each region's SALES numbers after RETURNS have been subtracted. Because SALES and RETURNS are by default ANALYSIS variables where statistic = sum, the *.statistic* suffix is included for derivations in the COMPUTE block (e.g. FINAMT=SALES.**SUM**-RETURNS.**SUM**).

❻ CUMFIN puts the total on every record. Though CUMFIN is not reported in the Final printed report, the variable is created as a report variable (versus a temporary variable) and kept in the outgoing SALES data set for reference.

❼ PCT derives each region's percentage of total sales by dividing each region's sales (minus returns) by the grand total (CUMFIN). Note that because CUMFIN is a COMPUTED variable rather than an ANAYSIS variable, the *.statistic* suffix is not included.

Table 8.3 shows the PROC REPORT output data set named Sales. The summary record (where _BREAK_ = _RBREAK_) has been kept here for the reader's reference.

Table 8.3 PROC REPORT Output Data Set SALES (Formatted With PROC PRINT)

Region	Sales	Returns	Finamt	cumfin	pct	_BREAK_
	$33,851,566	$1,172,092	$32,679,474	$32,679,474	100.0%	_RBREAK_
Africa	$2,342,588	$74,087	$2,268,501	$32,679,474	6.9%	
Asia	$460,231	$10,895	$449,336	$32,679,474	1.4%	
Canada	$4,255,712	$129,394	$4,126,318	$32,679,474	12.6%	
Central America/Caribbean	$3,657,753	$126,898	$3,530,855	$32,679,474	10.8%	
Eastern Europe	$2,394,940	$86,701	$2,308,239	$32,679,474	7.1%	
Middle East	$5,631,779	$206,880	$5,424,899	$32,679,474	16.6%	
Pacific	$2,296,794	$77,129	$2,219,665	$32,679,474	6.8%	
South America	$2,434,783	$102,851	$2,331,932	$32,679,474	7.1%	
United States	$5,503,986	$187,502	$5,316,484	$32,679,474	16.3%	
Western Europe	$4,873,000	$169,755	$4,703,245	$32,679,474	14.4%	

Obtain Regional Ranking Information

The fourth output in the ODS LAYOUT produces graphs for the "Top 3" regions. PROC RANK is used to easily attach a rank to each region in the SALES data set. Because the ranking variable (REGRANK) is needed in the SHOES data set (the source of the fourth output) the SALES REGRANK variable is merged back to SASHELP.SHOES BY REGION.

Code for Obtaining Regional Ranking Information

```
** Obtain Ranking of Region According to Sales; ❶
proc rank data=sales out=sales descending ties=low;
  var sales;
  ranks REGRANK; /* Names the rank variable REGRANK */
run;

** Merge Rank Information back to the Sales Data Set for Graph by
Product;
data shoes; ❷
  merge sales(keep= region regrank)
```

```
          sashelp.shoes(keep=region product sales returns);
   by region;
   if product in("Boot","Sandal","Sport Shoe") then product = "Other";
   else product=product;
run;
```

❶ PROC RANK is run on the SALES data set. Sales are ordered by DESCENDING value so the REGION with the highest sales obtains a rank of 1, the region with the second highest sales obtains a rank of 2, and so on. The new rank variable will be appended to the SALES data set and will be named REGRANK.

❷ SALES is merged back to SASHELP.SHOES to obtain the Regional Ranking variable. Table 8.4 displays the new data set, SHOES, after adding REGRANK.

Table 8.4 SHOES With Regional Rank (REGRANK) Added

Region	Sales	Returns	finamt	Cumfin	pct	_BREAK_	Regrank
Africa	$2,342,588	$74,087	$2,268,501	$32,679,474	6.9%		8
Asia	$460,231	$10,895	$449,336	$32,679,474	1.4%		10
Canada	$4,255,712	$129,394	$4,126,318	$32,679,474	12.6%		4
Central America/Caribbean	$3,657,753	$126,898	$3,530,855	$32,679,474	10.8%		5
Eastern Europe	$2,394,940	$86,701	$2,308,239	$32,679,474	7.1%		7
Middle East	$5,631,779	$206,880	$5,424,899	$32,679,474	16.6%		1
Pacific	$2,296,794	$77,129	$2,219,665	$32,679,474	6.8%		9
South America	$2,434,783	$102,851	$2,331,932	$32,679,474	7.1%		6
United States	$5,503,986	$187,502	$5,316,484	$32,679,474	16.3%		2
Western Europe	$4,873,000	$169,755	$4,703,245	$32,679,474	14.4%		3

Create a New ODS Style Template

Several modifications are made to the MEADOW template to style the SG graph elements and attributes in the ODS LAYOUT report.

```
proc template;
   Define style styles.MeadowG;  ❶
   Parent=styles.meadow;
```

```
          style graphborderlines / color=white;
          Style graphdata1 from graphdata1 / color=WHITE; /*SALE BARS*/ ❷
          Style graphdata2 from graphdata2 / color=RED;   /*RETURNS BARS*/ ❸
          style GraphFonts "Fonts used in graph styles" /
            "GraphDataFont" = ("Times New Roman",7pt) ❹
            "GraphUnicodeFont" = ("<MTsans-serif-unicode>",7pt)
            "GraphValueFont" = ("<sans-serif>, <MTsans-serif>",8pt,bold) ❺
            "GraphLabelFont" = ("Times New Roman,<sans-serif>,
                                <MTsans-serif>",10pt, bold) ❻
            "GraphLabel2Font" = ("<sans-serif>, <MTsans-serif>",7pt)
            "GraphFootnoteFont" = ("<sans-serif>, <MTsans-serif>",18pt)
            "GraphTitleFont" = ("<sans-serif>, <MTsans-serif>",11pt,bold) ❼
            "GraphTitle1Font" = ("<sans-serif>, <MTsans-serif>",8pt,bold)
            "GraphAnnoFont" = ("<sans-serif>, <MTsans-serif>",8pt);
       end;
       run;
```

❶ The MEADOW template is modified and saved as a new template named MEADOWG. Specific changes employed include modification of bar colors and various font attributes.

For the reader's reference, Figure 8.2 displays the "Shoe Sales and Returns by Region" bar chart without having applied the PROC TEMPLATE modifications. Figure 8.3 displays the same chart with the PROC TEMPLATE modifications.

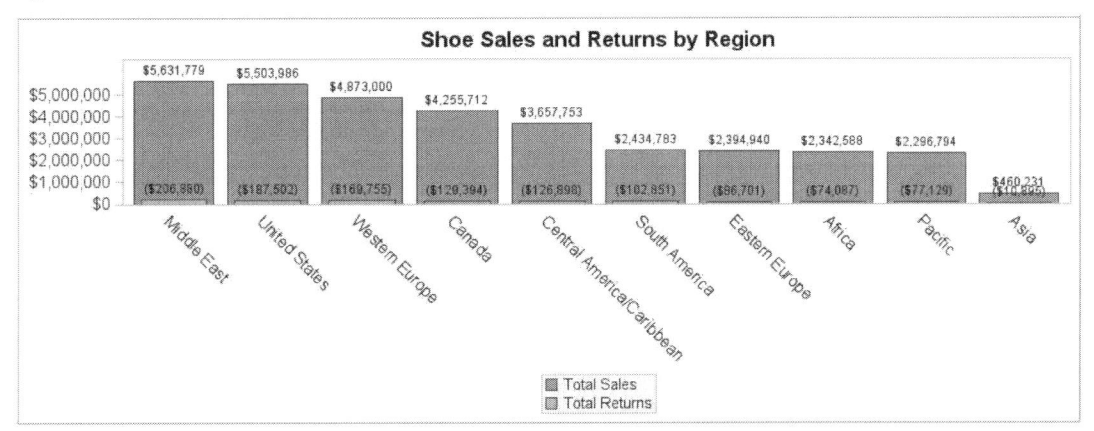

Chapter 8: Combining Graphs and Tabular Data **225**

Figure 8.2 WITHOUT PROC TEMPLATE Modifications

Figure 8.3 WITH PROC TEMPLATE Modifications

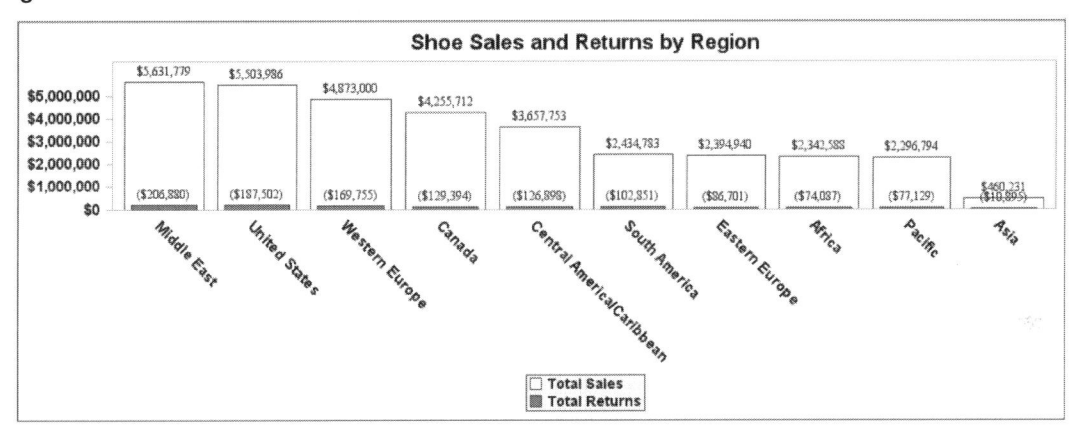

Table 8.5 lists some of the graph's elements and the portion of the graph they impact.

Table 8.5 Graph Style Elements

Graph Element	Applies to:
❷ GraphData1	First data grouping (e.g., Total Sales bars)
❸ GraphData2	Second data grouping (e.g., Total Returns bars)
❹ GraphDataFont	Data Labels (e.g., $5,631,779)
❺ GraphValueFont	X & Y Axis Data Point Labels (e.g., $0 tick mark label on y-axis, Middle East category label on x-axis, and legend items "Total Sales"

Graph Element	Applies to:
	and "Total Returns")
❻ GraphLabelFont	X & Y Axis Labels (Not shown here. E.g, if we had applied a Total Dollars label to the y-axis and a Region label to the x-axis, we could have modified the GraphLabelFont to our liking).
❼ GraphTitleFont	Graph Title (e.g., Shoe Sales and Returns by Region)

Create the ODS LAYOUT for the Report

This example uses a gridded ODS LAYOUT. We use the COLUMNS= and ROWS= options in the ODS LAYOUT START statement to divide the page into one column (COLUMNS=1) and four rows (ROWS=4). Each of the four rows will contain a graph or table. Four procedures are sandwiched between the ODS LAYOUT START statement and the ODS LAYOUT END statement.

```
** ODS PDF Specifications;
ods escapechar="^";
options nodate nonumber orientation=portrait;
ods pdf file="c:\temp\Ch8.pdf" bookmarklist=none style=MeadowG;

** Define Page Layout;
ODS LAYOUT START COLUMNS=1 COLUMN_WIDTHS=(7 IN) ROWS=4;

    {PROCEDURE 1 CODE}

    {PROCEDURE 2 CODE}

    {PROCEDURE 3 CODE}

    {PROCEDURE 4 CODE}

ODS LAYOUT END;
```

Create Formats Needed for Outputs

```
** Create Picture Formats;
proc format;
```

```
  **Dollar Format;  ❶

  picture retamt (round)

    low - high = "000,000,009)" (prefix="($");

  ** Percentage Format; ❷

  picture pctdec (round)

    0 - 1000 = "0009.9%"

    other    = " ";

run;
```

❶ The format RETAMT will allow us to apply parentheses to shoe return amounts in the vertical bar chart.

❷ The PCTDEC picture format will be used to apply the desired percentage format in the horizontal bar chart and the PROC REPORT table.

Use PROC SGPLOT to Create Vertical Bar Charts

PROC SGPLOT is used to overlay two vertical bar charts, one displaying Sales and the other displaying Returns.

Code for SGPLOT Vertical Bar Charts

```
ODS GRAPHICS / RESET imagename="Saleret1" imagefmt=PNG height=2.6 in
               width=7 in;  ❶
ODS LISTING GPATH = "c:\temp";  ❷
title "Shoe Sales and Returns by Region";
proc sgplot data=sales;
  vbar region / response=sales categoryorder=respdesc datalabel;  ❸
  vbar region / response=returns datalabel barwidth=.7;  ❹
  xaxis display=(nolabel noticks);  ❺
  yaxis display=(nolabel);  ❻
  format returns retamt.;  ❼
```

```
      keylegend / across=1;  ❽

run;
```

❶ Although ODS GRAPHICS is already enabled by default for SAS 9.3 SG procedures, we're using the ODS GRAPHICS statement to set ODS GRAPHICS options. The options set in this statement remain in effect for all graphics until we change or reset the settings with another ODS GRAPHICS statement. Table 8.6 further describes the settings we're using for the PROC SGPLOT figure.

Table 8.6 ODS GRAPHICS Specifications for the SGPLOT

Option	Description
RESET	RESET without a specification is the same as RESET=ALL. This resets all of the defaults.
IMAGENAME=	We name the image to be saved as SALERET1.
IMAGEFMT=	PNG is actually the default ODS GRAPHICS listing image format for SAS 9.3, but we specify the file type for clarity to the programmers.
HEIGHT= and WIDTH=	The graph dimensions are specified as 2.6 inches high and 7 inches wide.

❷ After the first ODS GRAPHICS statement, we tell SAS to save our images to the "c:\temp" folder with the statement ODS LISTING GPATH = "c:\temp". We do not reset this path, as want the images saved to this destination throughout this program.

PROC SGPLOT allows us to overlay the bar charts for SALES and RETURNS. The VBAR statements request Vertical Bar Charts.

❸ The first VBAR statement creates the outer vertical bar chart (SALES). We specify the category variable (the variable that immediately follows the word VBAR) as REGION. In doing so, each bar will represent a REGION.

The RESPONSE variable is specified as SALES, requesting that each region's bar height represents the value of SALES for that region.

CATEGORYORDER=RESPDESC orders our regions in **descending** order of the response variable, SALES. You can see the result is that the first region displayed is Middle East, as this has the highest sales. The following regions are displayed in order from highest to lowest sales, ending with the Asia region.

The DATALABEL option requests that the SALES amounts are shown above each bar.

❹ The second VBAR statement creates the inner vertical bar chart (RETURNS). Again, we specify the category variable as REGION.

The RESPONSE variable for this inner bar chart is RETURNS.

The DATALABEL option requests that the RETURNS amounts are shown above each bar.

The BARWIDTH option specifies the width of the bars as a ratio of the maximum possible width of 1 (the default is .8). Because we want the RETURNS bars to be narrower than the SALES bars, we specify the RETURNS BARWIDTHs to be .7.

❺ We suppress our X-axis label ("Region") and tick marks with the DISPLAY= (NOLABEL NOTICKS) specification.

❻ We suppress our Y-axis label.

❼ Here, the picture format we created for Return amounts is applied. The result is that we have the desired parentheses around these negative (in relation to sales) dollar amounts.

❽ The KEYLEGEND statement, along with the ACROSS option, is used to request that our legend contain only one column, i.e. as a stacked legend as shown in Figure 8.4. As shown in Figure 8.5, without this statement our graph shows the legend as one row with two columns.

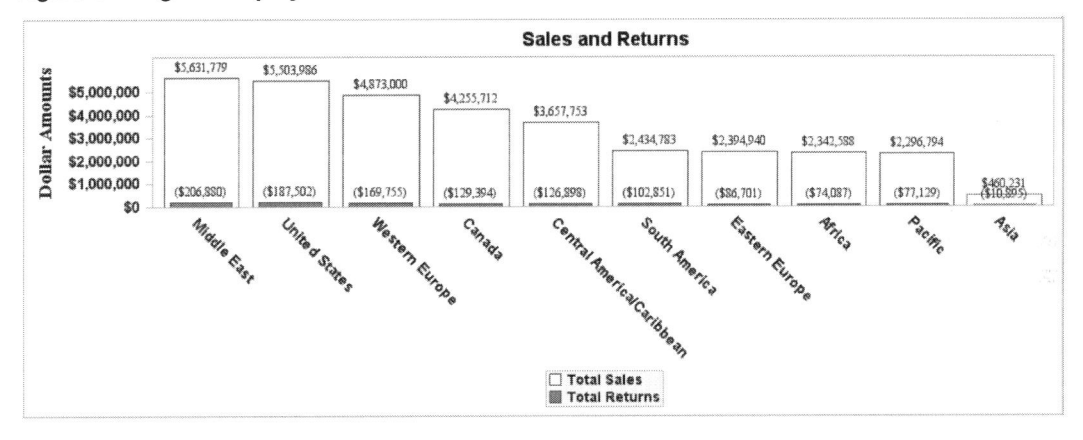

Figure 8.4 Legend Displayed as One Column

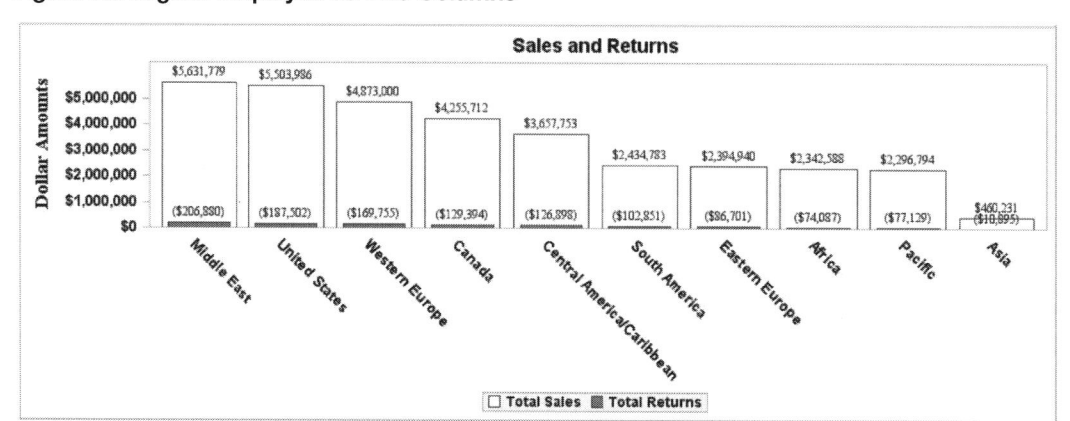

230 *PROC REPORT by Example: Techniques for Building Professional Reports Using SAS*

Figure 8.5 Legend Displayed as Two Columns

Using PROC SGPLOT to Create a Horizontal Bar Chart

A horizontal bar chart displaying each region's proportion of Total Sales is created with a separate PROC SGPLOT.

Horizontal Bar Chart Code

```
ODS GRAPHICS / RESET imagename="Distrib1" imagefmt=PNG height=2.5 in
               width=7 in;  ❶
proc SGPLOT data=SALES;
   title "Percentage of Total Shoe Sales (Minus Returns) by Region";
   hbar region / response=PCT categoryorder=respdesc datalabel
                 fillattrs=(color="verylightred");  ❷
   yaxis display=(nolabel);  ❸
   xaxis display=(nolabel);  ❹
   format pct pctdec.;  ❺
run;
```

❶ We specify the ODS GRAPHICS options that we want to apply to the horizontal bar chart.

❷   The HBAR statement requests a Horizontal Bar Chart. We specify the category variable (the variable that immediately follows the word HBAR) as REGION. In doing so, each bar will represent a REGION.

The RESPONSE variable is specified as PCT, the variable that contains each region's percentage of the total sales. The horizontal bar for each region will represent the value of PCT for that region.

CATEGORYORDER=RESPDESC orders the regions in descending order of the response variable, PCT. As with the vertical bar chart, the result is that the first region displayed is Middle East, as this has the highest PCT. The following regions are displayed in order from highest to lowest PCT, ending with the Asia region.

The DATALABEL option requests that the PCT values are shown to the right of each horizontal bar.

The FILLATTRS= option specifies style elements for the bar fill. We are changing the fill color to very light red.

❸   We suppress the X-axis label.
❹   Likewise, we suppress the Y-axis label.
❺   The picture format PCTDEC is applied. The result is that we have a rounded percentage value displayed to one decimal place with the "%" sign attached.

Using PROC REPORT to Obtain Tabular Output

Our third row of output in the ODS Layout is a table. As with the previous two outputs, the summary data set SALES is the source for this report.

```
proc report data=sales nowd split="|"

  style(report)=[outputwidth=7 in]

  style(column)=[just=c];  ❶

  column sales=SALESORD region sales returns finamt pct;  ❷

  ** DEFINE Specifications;  ❸

  define SALESORD /  order order=internal descending noprint;

  define region   /  "Region" style(column)=[just=left];

  define sales    /  "Sales";

  define returns  /  "Returns";
```

232 *PROC REPORT by Example: Techniques for Building Professional Reports Using SAS*

```
    define finamt    /  "Sales Minus|Returns" format=dollar10.;
    define pct       /  "Percentage|of Total" format=pctdec.;
run;
```

❶ The output width is set to 7 inches so the table fills the full ODS Layout column.

❷ We create an alias for SALES (named SALESORD) so we can order rows by SALES before ordering by REGION. (Recall that columns are processed from left to right. We need the SALES column or an alias to be listed and DEFINED as ORDER **prior** to REGION). We apply the NOPRINT option to SALESORD and print the SALES column listed after REGION, reflecting the desired order of the SALES column in the printed report.

❸ This is a simple PROC REPORT for which the incoming data only has one record per region and no statistics are performed on the numeric variables, which would by default be analysis variables. SALES was already summed by REGION in the first PROC REPORT and we want to use it as an ORDER variable for the final report. Though not specified as ORDER in the DEFINE statement, SALES is defined as ORDER by default since its alias SALESORD was defined as an ORDER variable earlier.

Using PROC SGPANEL to Create Bar Charts for the Top 3 Regions

Our final graph uses PROC SGPANEL to obtain vertical bar charts for the three regions ranked the highest in sales. PROC SGPANEL allows us to create three side-by-side charts with one procedure. These charts use the SHOES data set as the input source, and each bar chart displays Sales by Product. Because SGPANEL is set up to produce panels, we do not need to specify additional ODS LAYOUT options.

```
ODS GRAPHICS / RESET imagename="top31" imagefmt=PNG width=7 in
               height=2.4 in;  ❶

title height=8 pt " ";

title2 height=11 pt "Top 3 Regions: Shoe Sales by Product";

title3 height=8 pt    " ";

proc sgpanel data=shoes(where=(regrank in(1,2,3)));  ❷
   panelby region / layout=columnlattice spacing=48 novarname;  ❸
   vbar product / response=sales datalabel categoryorder=respdesc
       datalabelattrs=(size=6 pt) fillattrs=(color="verylightred");  ❹
```

Chapter 8: Combining Graphs and Tabular Data **233**

```
  ** Remove Axis Labels to Save Space;
  rowaxis display=(nolabel);
  colaxis display=(nolabel);
run;

** END OF ODS LAYOUT;  ❺
ODS LAYOUT END;
ods pdf close;
title;
```

❶ We specify the ODS GRAPHICS options we want to apply to the "Top 3" bar charts.

❷ We are now using PROC SGPANEL instead of PROC SGPLOT. We run PROC SGPANEL, only on data for which a region ranks in the top 3 in terms of sales (based on REGRANK from the PROC RANK procedure).

❸ The PANELBY statement controls our layout. If we specified the statement as PANELBY REGION without any options, the default graph would appear as in Figure 8.6.

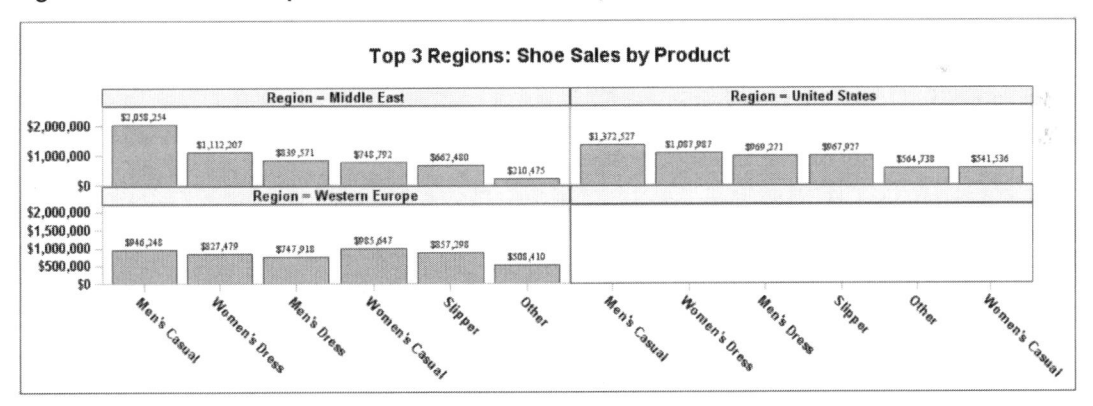

Figure 8.6 SGPANEL Graph With Default PANELBY Specifications

In order to achieve the three side-by-side panels with the desired spacing and borders, we apply the following options:

- LAYOUT=COLUMNLATTICE gives us a panel, or column for each of the three regions.
- SPACING=48 provides the desired spacing between each panel (e.g., between the Middle East chart and United States chart, and between the United States chart and Western Europe chart).

- The NOVARNAME option removes the "Region = " text from the chart headers.

The application of these options results in the desired graph, as shown in Figure 8.7.

Figure 8.7 SGPANEL Graph with PANELBY Specifications Applied

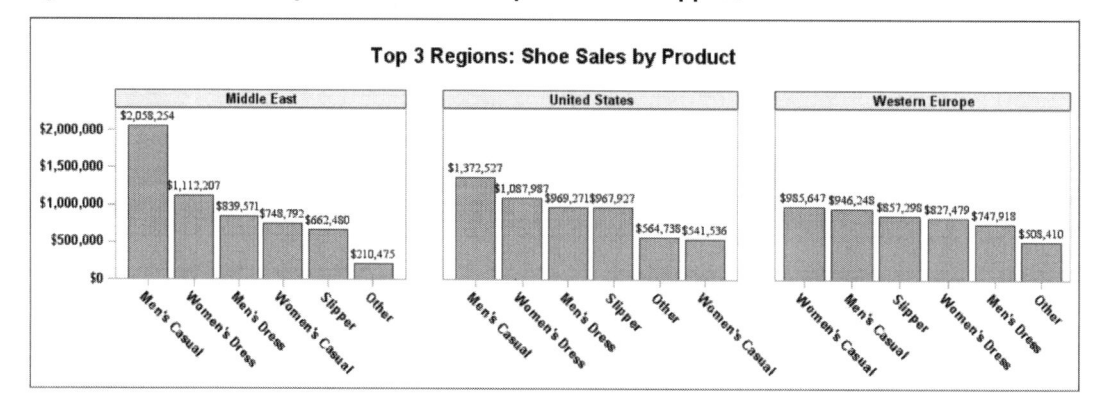

❹ The VBAR statement creates the vertical bar charts for sales. We specify the category variable as PRODUCT.

The RESPONSE variable for the bar chart is SALES. (We do not take RETURNS into account in this by PRODUCT chart).

The DATALABEL option requests that the sales amounts are shown above each bar.

CATEGORYORDER=RESPDESC orders our products in descending order of the response variable, SALES.

We use the DATALABELATTRS= option to set the data labels to 6 pt. font size.

We specify the fill attributes of the bars to be very light red with the FILLATTRS= option.

❺ The (required) ODS LAYOUT END statement closes this ODS layout.

Chapter 8 Summary

This chapter showed how to:

Create a New Data Set with PROC REPORT	The new data set named SALES summarized Region Total Sales, Total Returns, and Percent of Total.
Obtain Region Rank Information	PROC RANK was used to attach a rank to each region in the SALES data. The RANK information was merged with SASHELP.SHOES to put regional ranking on every record.
Create a New ODS Style Template	Several modifications were made to the MEADOW template to style the graph elements and attributes in the report.
Create the ODS LAYOUT for the Report	The key statements around the procedures were ODS LAYOUT START and ODS LAYOUT END. The ROWS= and COLUMNS= options in this statement were used to set up the page grid.
Use PROC SGPLOT to Create Vertical Bar Charts	PROC SGPLOT was used to overlay two vertical bar charts, one displaying Sales and the other displaying Returns.
Use PROC SGPLOT to Create a Horizontal Bar Chart	A horizontal bar chart displaying each region's proportion of Total Sales was also created with a separate PROC SGPLOT.
Use PROC REPORT to Obtain Tabular Output	This table also used the new SALES data set to produce the information in tabular format.
Use PROC SGPANEL to Create Bar Charts for Top 3 Regions	The three highest ranking regions in terms of sales displayed vertical bar charts based on the SHOES data. PROC SGPANEL allowed us to create the three graphs with one procedure and display them side by side.

Chapter 9: Using PROC REPORT to Obtain Summary Statistics for Comparison

Introduction 238

Example: Vehicle MSRP Comparison Report 238

Goals for MSRP Comparison Report 240
    Key Steps 240

Source Data 240

ODS Style Template Used 242

Programs Used 242

Implementation 242

Initial PROC REPORT for Obtaining Statistics 242
    Code for Obtaining Statistics 242

Produce the Report 245
    Code for Print Report 245

Chapter 9 Summary 254

Introduction

A number of summary statistics can be obtained with the REPORT procedure. To mention just a few, we can obtain counts, percentages, means, standard deviations, medians (50th percentile), 25th and 75th percentiles, and minimum and maximum values. Once we have these statistics, additional analyses such as comparisons of individual records to summary statistics can be performed using COMPUTE blocks.

Example: Vehicle MSRP Comparison Report

A report is produced to summarize Manufacturer's Suggested Retail Price (MSRP) for vehicles by continent of origin (Asia, Europe, and USA) and vehicle type (e.g., SUV, Sedan, Sport). Statistics including quartiles and minimum and maximum MSRP for each Continent-Vehicle Type group are obtained. Within each continent and vehicle type, individual vehicle MRSPs are compared to the group statistics to determine which pricing category the vehicle falls within (e.g., which percentile). Specific report features include:

- Above each Continent-Vehicle Type table, the percentiles and the lowest (minimum) and highest (maximum) MSRP are reported.
- The individual vehicles that represent the highest and lowest priced vehicles within Origin-VehicleType are highlighted (shaded) within the table cells (see "MSRP" column).
- A report column, titled "MSRP Price Point" displays $ symbols to express the price rating of each vehicle, with a single "$" representing the lowest priced vehicles (25th percentile) and "$$$$" representing the highest priced vehicles (> 75th percentile).

Figure 9.1 displays an example page of the report.

Chapter 9: Using PROC REPORT to Obtain Summary Statistics for Comparison **239**

Figure 9.1 Partial Print of MSRP Report

Continent of Origin: USA
Vehicle Type: SUV

SUV MSRP Price Point

MSRP <=25th Percentile ($26,545) ($)
MSRP <=50th Percentile ($32,235) ($$)
MSRP <=75th Percentile ($42,735) ($$$)
MSRP > 75th Percentile ($42,735) ($$$$)

■ Lowest MSRP: $20,130
■ Highest MSRP: $52,795

Make	Model	Cylinders	Horsepower	MSRP	MSRP Price Point
Buick	Rendezvous CX	6	185	$26,545	$
	Rainier	6	275	$37,895	$$$
Cadillac	SRX V8	8	320	$46,995	$$$$
	Escalade	8	295	$52,795	$$$$
Chevrolet	Tracker	6	165	$20,255	$
	TrailBlazer LT	6	275	$30,295	$$
	Tahoe LT	8	295	$41,465	$$$
	Suburban 1500 LT	8	295	$42,735	$$$
Dodge	Durango SLT	8	230	$32,235	$$
Ford	Escape XLS	6	201	$22,515	$
	Explorer XLT V6	6	210	$29,670	$$
	Expedition 4.6 XLT	8	232	$34,560	$$$
	Excursion 6.8 XLT	10	310	$41,475	$$$
GMC	Envoy XUV SLE	6	275	$31,890	$$
	Yukon 1500 SLE	8	285	$35,725	$$$
	Yukon XL 2500 SLT	8	325	$46,265	$$$$
Hummer	H2	8	316	$49,995	$$$$
Jeep	Liberty Sport	4	150	$20,130	$
	Wrangler Sahara convertible 2dr	6	190	$25,520	$
	Grand Cherokee Laredo	6	195	$27,905	$$
Lincoln	Aviator Ultimate	8	302	$42,915	$$$$
	Navigator Luxury	8	300	$52,775	$$$$
Mercury	Mountaineer	6	210	$29,995	$$
Pontiac	Aztekt	6	185	$21,595	$
Saturn	VUE	4	143	$20,585	$

Goals for MSRP Comparison Report

The vehicle report uses behind-the-scenes steps to determine each vehicle's MSRP percentile category, as well as the minimum and maximum values. By "behind-the-scenes" we mean that these statistics are not printed in columns. They are used in COMPUTE blocks for comparison and are reported as summary information above the report for each vehicle type and as symbols and highlighted cells within columns.

Key Steps

The REPORT procedure is run twice, with the first run performed simply for the purpose of obtaining a data set with needed statistics. This summary data set is merged back to the full data set so that comparisons to percentiles and the minimum and maximum MSRPs can be made on a record-by-record basis.

The second PROC REPORT produces the printed report. Several PROC REPORT options are used, including:

- The use of BY VARIABLES and placement of BY values in page titles
- The SPANROWS option for ORDER variables
- ALIASES for computing new variables and ordering rows
- Table of Contents options

Source Data

The source data set is the SAS supplied data set SASHELP.CARS (2004 Car Data). Only the variables needed for this report are kept. Table 9.1 shows a partial print of the data, and Table 9.2 displays the variable attributes of the data set.

Table 9.1 Partial Print of CARS Data

Make	Model	Type	Origin	MSRP	Cylinders	Horsepower
Buick	Rainier	SUV	USA	$37,895	6	275
Buick	Rendezvous CX	SUV	USA	$26,545	6	185
Cadillac	Escalade	SUV	USA	$52,795	8	295
Cadillac	SRX V8	SUV	USA	$46,995	8	320
Chevrolet	Suburban 1500 LT	SUV	USA	$42,735	8	295
Chevrolet	Tahoe LT	SUV	USA	$41,465	8	295
Chevrolet	TrailBlazer LT	SUV	USA	$30,295	6	275
Chevrolet	Tracker	SUV	USA	$20,255	6	165
Dodge	Durango SLT	SUV	USA	$32,235	8	230
Ford	Excursion 6.8 XLT	SUV	USA	$41,475	10	310
Ford	Expedition 4.6 XLT	SUV	USA	$34,560	8	232
Ford	Explorer XLT V6	SUV	USA	$29,670	6	210
Ford	Escape XLS	SUV	USA	$22,515	6	201
GMC	Envoy XUV SLE	SUV	USA	$31,890	6	275
GMC	Yukon 1500 SLE	SUV	USA	$35,725	8	285
GMC	Yukon XL 2500 SLT	SUV	USA	$46,265	8	325
Hummer	H2	SUV	USA	$49,995	8	316
Jeep	Grand Cherokee Laredo	SUV	USA	$27,905	6	195
Jeep	Liberty Sport	SUV	USA	$20,130	4	150
Jeep	Wrangler Sahara convertible 2dr	SUV	USA	$25,520	6	190
Lincoln	Navigator Luxury	SUV	USA	$52,775	8	300
Lincoln	Aviator Ultimate	SUV	USA	$42,915	8	302
Mercury	Mountaineer	SUV	USA	$29,995	6	210
Pontiac	Aztekt	SUV	USA	$21,595	6	185
Saturn	VUE	SUV	USA	$20,585	4	143
Buick	Century Custom 4dr	Sedan	USA	$22,180	6	175
Buick	LeSabre Custom 4dr	Sedan	USA	$26,470	6	205
Buick	Regal LS 4dr	Sedan	USA	$24,895	6	200
Buick	Regal GS 4dr	Sedan	USA	$28,345	6	240

Table 9.2 Contents of CARS Data

#	Variable	Type	Len	Format
1	Make	Char	13	
2	Model	Char	40	
3	Type	Char	8	
4	Origin	Char	6	
5	MSRP	Num	8	DOLLAR8.
6	Cylinders	Num	8	
7	Horsepower	Num	8	

ODS Style Template Used

HARVEST is the ODS Style template used to produce Figure 9.1.

Programs Used

The name of the program used is Ch9Stat.sas.

Implementation

Initial PROC REPORT for Obtaining Statistics

The purpose of the first PROC REPORT is to obtain percentile statistics (25th percentile, median, and 75th percentile) and the minimum and maximum values for MSRP by continent of origin and by vehicle type. The statistics are saved to a data set named QUARTILES. This PROC REPORT data set is merged back to the original data set so that comparisons to percentiles and minimum and maximum values can be made on a record-by-record basis.

Code for Obtaining Statistics

```
proc sort data=sashelp.cars(keep=Make Model Type Origin MSRP Cylinders
                   Horsepower)
     OUT=CARS;
```

Chapter 9: Using PROC REPORT to Obtain Summary Statistics for Comparison   243

```
  by origin type;  ❶
run;

proc report data=cars nowd OUT=QUARTILES;  ❷
  column origin type msrp msrp=msrp2 msrp=msrp3 msrp=msrp4 msrp=msrp5;
❸
  define origin  / group;
  define type    / group;
  ** Define Statistics;  ❹
  define msrp   / p25;
  define msrp2  / median;
  define msrp3  / p75;
  define msrp4  / min;
  define msrp5  / max;
run;

data cars;  ❺
  merge cars
        quartiles(rename=(msrp=per25 msrp2=per50 msrp3=per75
                          msrp4=pmin msrp5=pmax));
  by origin type;
run;
```

❶ The SAS data set SASHELP.CARS is sorted by ORIGIN and TYPE and the new sorted data set is named CARS. The ORIGIN and TYPE sort is needed for a later merge.

❷ PROC REPORT is used to create a data set which contains the MSRP percentiles. The data set is named QUARTILES with the OUT= option.

❸ Note that the incoming variable MSRP is used for five PROC REPORT columns. Four aliases, MSRP2 through MSRP5, are created for MSRP so the variable can be the source for five different statistics in the DEFINE statements that follow.

❹ The DEFINE statement for MSRP requests the 25th Percentile statistic amount with "P25".

In the following DEFINE statements,

- MSRP2 requests the Median
- MSRP3 requests the 75th Percentile (with "P75")
- MSRP4 requests the minimum MSRP (with min)
- MSRP5 requests the maximum MSRP (with max)

A print of the new data set QUARTILES is shown in Table 9.3.

Table 9.3 Partial Print (WHERE ORIGIN="USA") of PROC REPORT Output Data Set QUARTILES

Origin	Type	MSRP	msrp2	msrp3	msrp4	msrp5	_BREAK_
USA	SUV	$26,545	$32,235	$42,735	$20,130	$52,795	
USA	Sedan	$19,090	$24,260	$30,835	$10,995	$50,595	
USA	Sports	$33,500	$37,530	$51,535	$18,345	$81,795	
USA	Truck	$19,488	$23,703	$34,820	$14,385	$52,975	
USA	Wagon	$17,475	$22,290	$23,560	$17,045	$31,230	

❺ The QUARTILES data set is merged back to CARS. The MSRP variables are renamed to have more meaningful names that describe the percentiles they represent. Table 9.4 displays the merged data set.

Table 9.4 Partial Print of Merged Data Set (some variables excluded)

Make	Model	Type	Origin	MSRP	per25	per50	per75	pmin	pmax
Buick	Rainier	SUV	USA	$37,895	$26,545	$32,235	$42,735	$20,130	$52,795
Buick	Rendezvous CX	SUV	USA	$26,545	$26,545	$32,235	$42,735	$20,130	$52,795
Cadillac	Escalade	SUV	USA	$52,795	$26,545	$32,235	$42,735	$20,130	$52,795
Cadillac	SRX V8	SUV	USA	$46,995	$26,545	$32,235	$42,735	$20,130	$52,795
Chevrolet	Suburban 1500 LT	SUV	USA	$42,735	$26,545	$32,235	$42,735	$20,130	$52,795
Chevrolet	Tahoe LT	SUV	USA	$41,465	$26,545	$32,235	$42,735	$20,130	$52,795
Chevrolet	TrailBlazer LT	SUV	USA	$30,295	$26,545	$32,235	$42,735	$20,130	$52,795
Chevrolet	Tracker	SUV	USA	$20,255	$26,545	$32,235	$42,735	$20,130	$52,795
Dodge	Durango SLT	SUV	USA	$32,235	$26,545	$32,235	$42,735	$20,130	$52,795
Ford	Excursion 6.8 XLT	SUV	USA	$41,475	$26,545	$32,235	$42,735	$20,130	$52,795

Make	Model	Type	Origin	MSRP	per25	per50	per75	pmin	pmax
Ford	Expedition 4.6 XLT	SUV	USA	$34,560	$26,545	$32,235	$42,735	$20,130	$52,795
Ford	Explorer XLT V6	SUV	USA	$29,670	$26,545	$32,235	$42,735	$20,130	$52,795
Ford	Escape XLS	SUV	USA	$22,515	$26,545	$32,235	$42,735	$20,130	$52,795
GMC	Envoy XUV SLE	SUV	USA	$31,890	$26,545	$32,235	$42,735	$20,130	$52,795
GMC	Yukon 1500 SLE	SUV	USA	$35,725	$26,545	$32,235	$42,735	$20,130	$52,795
GMC	Yukon XL 2500 SLT	SUV	USA	$46,265	$26,545	$32,235	$42,735	$20,130	$52,795
Hummer	H2	SUV	USA	$49,995	$26,545	$32,235	$42,735	$20,130	$52,795
Jeep	Grand Cherokee Laredo	SUV	USA	$27,905	$26,545	$32,235	$42,735	$20,130	$52,795

Produce the Report

Now that we have the group statistics merged back to the CARS data, we are ready to produce the print report.

Code for Print Report

```
** Titles; ❶

TITLE  "Continent of #byvar1: #byval(origin)";

TITLE2 "Vehicle #byvar2: #byval(type)";

** ODS PDF Specifications;

ods escapechar="^";

options nobyline nodate nonumber orientation=portrait; ❷

ods _all_ close;

ods pdf file = "c:\temp\Ch10Cars.pdf"
    uniform pdftoc=2 style=harvest; ❸
```

246 *PROC REPORT by Example: Techniques for Building Professional Reports Using SAS*

```
ods proclabel="MSRP Report by Origin and Type";  ❹


proc report data=cars nowd spanrows split="|" missing
   style(report)=[asis=on];  ❺

   by origin type;  ❻

   column type=type2 make msrp=MSRPORD model cylinders horsepower msrp
          per25 per50 per75 pmin pmax msrpptle;  ❼

   ** DEFINE Specifications;  ❽

   define type2     /  noprint;
   define make      / order style(column)=[font_weight=bold];
   define MSRPORD / order noprint;
   define model     / order;
   define cylinders   / order style(column)=[just=c];
   define horsepower / order style(column)=[just=c];

   define per25 / noprint;
   define per50 / noprint;
   define per75 / noprint;
   define pmin  / noprint;
   define pmax  / noprint;

   define msrpptle / computed "MSRP|Price Point"
                     style(column)=[just=l indent=.75 in
                                    cellwidth=1.8 in];

   ** Create Price Symbols Column and Highlight Min and Max Rows;  ❾
   compute msrpptle / char length=6;

      ** Determine Percentile and Assign $ Symbols;
      if msrp.sum <= per25.sum then msrpptle="$";
```

```
      else if per25.sum < msrp.sum <=per50.sum then msrpptle="$$";

      else if per50.sum < msrp.sum<= per75.sum then msrpptle="$$$";

      else if msrp.sum > per75.sum then msrpptle="$$$$";

   ** Color Min and Max Cells;

   if pmin.sum=msrp.sum then

      call define('msrp.sum','style','style={background=blue
                foreground=white font_weight=bold}');

   if pmax.sum=msrp.sum then

      call define('msrp.sum','style','style={background=red
                foreground=white font_weight=bold}');

  endcomp;

compute before _PAGE_ / left;  ❿

length text0 - text6 $100;

if _BREAK_=' ' then

  do;

     text0="^{style [font_size=12 pt textdecoration=underline]"
           ||strip(type2)||" MSRP Price Point}";

     text1="MSRP <=25th Percentile ("
           ||strip(put(per25.sum,dollar10.))||") ($)";

     text2="MSRP <=50th Percentile ("
           ||strip(put(per50.sum,dollar10.))||") ($$)";

     text3="MSRP <=75th Percentile ("
           ||strip(put(per75.sum,dollar10.))||") ($$$)";

     text4="MSRP > 75th Percentile ("
           ||strip(put(per75.sum,dollar10.))||") ($$$$)";

     ** Min/Max Legend with text;
```

```
      text5=
       "^{style [font_face=wingdings font_size=12 pt foreground=blue]n}"
           ||"Lowest  MSRP: " || strip(put(pmin.sum,dollar10.));
      text6=
       "^{style [font_face=wingdings font_size=12 pt foreground=red]n}"
           ||"Highest MSRP: " || strip(put(pmax.sum,dollar10.));
    end;
  ** Put New Variables in Line Statements;
    line text0 $100.;
    line '';
    line text1 $100.;
    line text2 $100.;
    line text3 $100.;
    line text4 $100.;
    line '';
    line text5 $100.;
    line text6 $100.;
  endcomp;
  compute before make;
    line '';
  endcomp;
  run;
  ods pdf close;
  ods html;
```

❶ ORIGIN is the first BY variable. #BYVAR1 provides the name of this variable, so "Continent of #BYVAR1" translates into "Continent of Origin." Of course, we could have just typed the word "Origin," but for this example we are demonstrating this feature.

#BYVAL(*variable*) provides the value of the variable specified in parentheses, allowing a dynamic title for each value of the BY variable. For example, when the Origin is USA, #BYVAL(ORIGIN) translates into "USA" in the page title.

TYPE is the second BY variable. #BYVAR2 provides the name of this variable, so Vehicle #BYVAR2 translates into "Vehicle Type."

#BYVAL(TYPE) translates into "SUV," "Sedan," and other vehicle types depending on the page of the report.

❷ NOBYLINE: The report tables are produced by ORIGIN and by TYPE. The NOBYLINE option is specified so we can customize our own format of the "BY LINEs," which we will insert into the page titles.

❸ ODS PDF Specifications

The UNIFORM option keeps the BY group tables the same width.

Table of Contents Specifications

A table of contents (TOC) is produced by default in the PDF destination. It can be omitted with the NOTOC option. The TOC does not show on the printed report, but it is available onscreen so the user can select the portion of the report they would like to view. For the MSRP report, we want to keep the table of contents and change some of the default TOC settings.

Specifically, we want a user to easily know which link to click to get to a desired section of the report. Figure 9.2 displays the default TOC as it appears on page 1 of the report.

Using Asia Hybrid as an example, note that there are currently 4 nodes in the default TOC.

- The Report Procedure
- Origin=Asia Type=Hybrid
- Detailed and/or summarized report
- Table 1

Clicking on either of the last two nodes ("Detailed and/or summarized report" or "Table 1") does not provide additional functionality for this report; both nodes lead to the Origin=Asia Type=Hybrid report. Since these last two nodes are extraneous, we would like to remove them.

We do this by setting the TOC level of node expansion to 2 with the code **PDFTOC=2**. The result is shown in Figure 9.3, for which we now only see the first two nodes ("The Report Procedure" and "Origin=Asia Type=Hybrid").

250 PROC REPORT by Example: Techniques for Building Professional Reports Using SAS

Figure 9.2 Default TOC

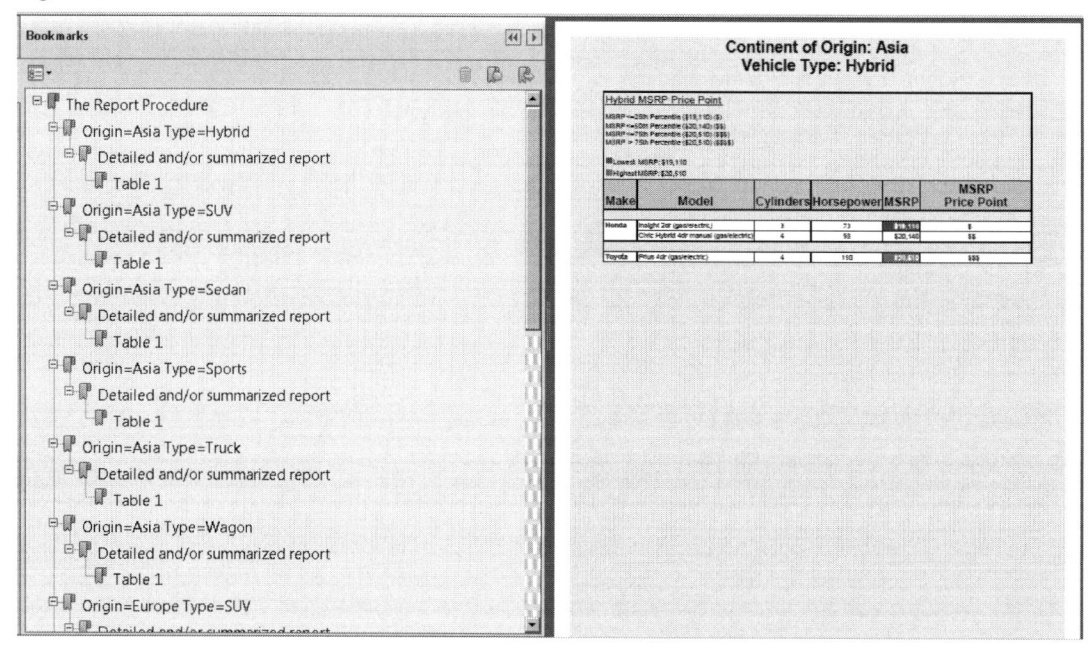

Figure 9.3 TOC with Revised Node Expansion (Reduction)

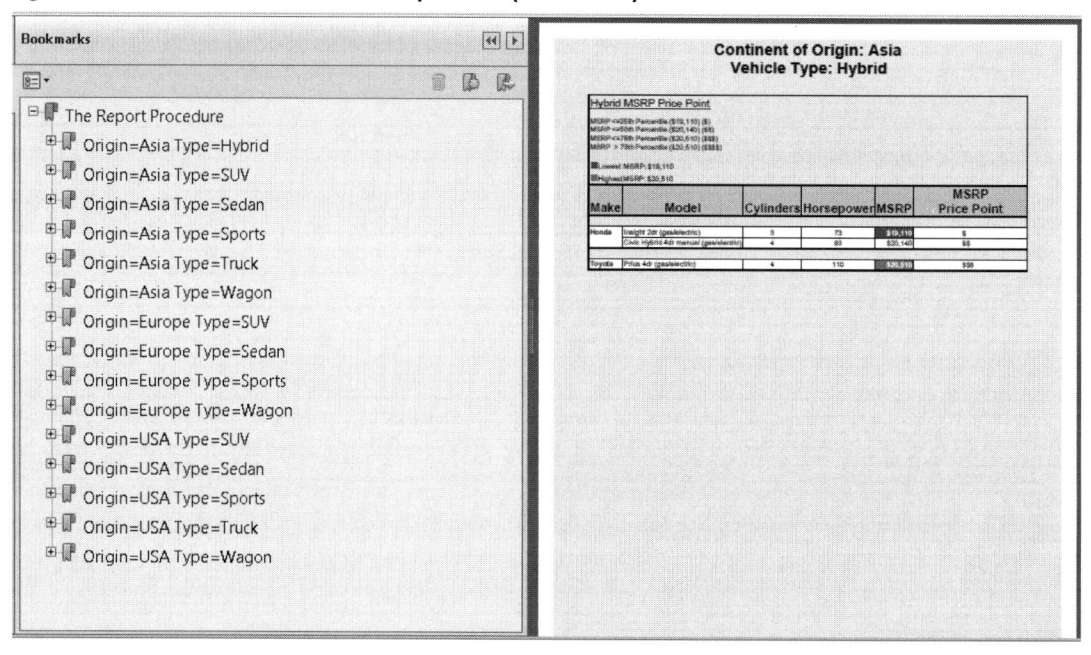

Chapter 9: Using PROC REPORT to Obtain Summary Statistics for Comparison **251**

❹ With the ODS PROCLABEL (procedure label) option, we are able to further change the TOC Appearance by Changing the first node's text from "The Report Procedure" to "MSRP Report by Origin and Type." Figure 9.4 shows the final TOC.

Figure 9.4 TOC with Procedure Label Changed

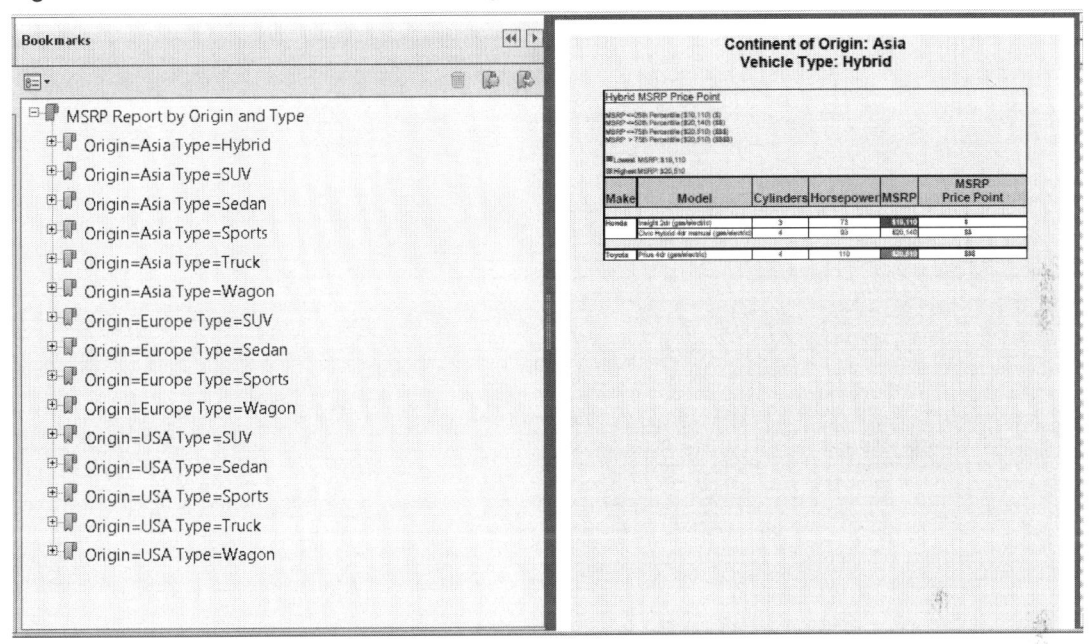

❺ SPANROWS, an option added with SAS 9.2 is used to create a single cell for each level of vehicle type. Note how the "Buick" cell in Figure 9.1 spans across both "Model" rows. ASIS=ON is used to preserve leading spaces in text that we have throughout the report.

❻ The BY statement specifies that tables should be produced by ORIGIN and TYPE.

❼ Aliases TYPE2 and MSRPORD are created.

TYPE2 is needed for the COMPUTEd TEXT0 variable in which we insert the vehicle TYPE before each page (example, "SUV MSRP Price Point").

MSRPORD is needed for ordering rows.

The desired row order is MAKE, MSRP, and then MODEL. However, we want to display MSRP after MODEL as the fifth column in the report.

The following COLUMN statement leads to the output in Figure 9.5, in which Models are in alphabetical order rather than the desired order of ascending MSRP within Make.

```
column MAKE MODEL cylinders horsepower MSRP
       per25 per50 per75 pmin pmax msrpptle;
```

This occurs because report variables are processed from left to right, therefore the report in Figure 9.5 is ordered by make and model first, and later by MSRP.

Figure 9.5 Rows Not in Desired Ascending MSRP Order

Continent of Origin: USA
Vehicle Type: SUV

SUV MSRP Price Point

MSRP <=25th Percentile ($26,545) ($)
MSRP <=50th Percentile ($32,235) ($$)
MSRP <=75th Percentile ($42,735) ($$$)
MSRP > 75th Percentile ($42,735) ($$$$)

■ Lowest MSRP: $20,130
■ Highest MSRP: $52,795

Make	Model	Cylinders	Horsepower	MSRP	MSRP Price Point
Buick	Rainier	6	275	$37,895	$$$
	Rendezvous CX	6	185	$26,545	$
Cadillac	Escalade	8	295	$52,795	$$$$
	SRX V8	8	320	$46,995	$$$$

To obtain the desired row order, we create MSRPORD to be used as an ORDER variable before MODEL, and suppress the printing of MSRPORD with NOPRINT. Later in the COLUMN statement, MSRP is listed for printing. The final COLUMN statement is specified as:

```
column type=type2 make msrp=MSRPORD model cylinders horsepower msrp
       per25 per50 per75 pmin pmax msrpptle;
```

❽ The only printed columns are MAKE, MODEL, CYLINDERS, HORSEPOWER, MSRP, and the COMPUTEd column MSRPPTLE. The other report variables are used for other purposes and are specified as NOPRINT in the DEFINE statements.

❾ The following steps are taken to obtain the MSRP Price Point "$" symbols.

- A new COMPUTE variable, MSRPPTLE is created. It is specified as a character (CHAR) variable. The character (or char) designation is necessary for computed character variables. The length is specified as 6.

- IF and ELSE IF statements are used to determine into which percentile each record's MSRP falls. Because MSRP and the percentile variables are ANALYSIS variables, the .SUM suffix is needed for the COMPUTE block to recognize the variables. MSRPPTLE $ values are assigned according to each vehicle's percentile placement.

Call DEFINE is used to change an MSRP cell's background color to blue and font color to white, if its MSRP.SUM value equals the Minimum MSRP.

Likewise, Call DEFINE is used to change an MSRP cell's background color to red and font color to white, if its MSRP.SUM value equals the Maximum MSRP.

⑩ COMPUTE BEFORE _PAGE_ causes the compute block to execute once for each page after printing the titles. LEFT is specified so all of the LINEs specified in this block will be left justified.

Seven text variables (TEXT0 through TEXT6) that will be placed in corresponding line statements are created.

- TEXT0 contains the main header, for example: "SUV MSRP Price Point"
    - The inline formatting function STYLE (used along with our declared ODS character "^") allows us to style our header to be 12 point font and underlined.
    - The STYLE function, enclosed in {}, has two arguments:
        - our style overrides, enclosed in [ ] and,
        - the text to be formatted (the text of our TYPE2 variable concatenated with "MSRP Price Point").

- TEXT1 through TEXT4 variables contain the MSRP Price Point descriptions.
    - For example, "MSRP<=25th Percentile ($26,545) ($)".

- TEXT5 and TEXT6 variables contain the Minimum and Maximum values, respectively. A colored symbol is placed to the left of each of these to serve as a legend for the colored minimum and maximum cells in the table body.
    - Style function are used to apply Wingdings font to the 'n' character to display this as a square.
    - The color of the square for lowest MSRP is applied a font color of blue. The square for the highest MSRP is given a red foreground.

254 *PROC REPORT by Example: Techniques for Building Professional Reports Using SAS*

- The square is concatenated to the description (Highest or Lowest) along with the corresponding minimum or maximum value.

TEXT0 through TEXT6 are put in individual line statements to be placed before each page of the report.

Chapter 9 Summary

This chapter covered how to create the MSRP report using the following steps.

- PROC REPORT was used to easily obtain group statistics, including quartiles and minimum and maximum values.
- The PROC REPORT output data set was merged back to the CARS data to allow for easy comparison of individual vehicle MSRPs to the group statistics.
- A second PROC REPORT created the printed report. Some of the items this section demonstrated included:
    - how to modify a PDF Table of Contents with the PDFTOC option and ODS PROCLABEL statement
    - the use of #BYVAR and #BYVAL options within a title statement
    - the use of an alias for BY variables so that they could be used in a COMPUTE block
    - the use of an alias to order rows
    - the SPANROWS option which allows a group or order variable to display in one cell that spans across the individual rows in that grouping
    - how to insert text lines and symbols above each page of a report

References

Allison, Robert. "Robert Allison's SAS/Graph Examples!" Available at http://robslink.com/SAS/Home.htm.

Andrews, Rick. 2011. "Printable Spreadsheets Made Easy: Utilizing the SAS® Excel XP Tagset." SESUG 2011: The Proceedings of the SouthEast SAS Users Group - Paper RV-05. Cary, NC: SAS Institute Inc. Available at http://analytics.ncsu.edu/sesug/2011/RV05.Andrews.pdf.

Bessler, LeRoy. 2012. "Comparison of SAS® Graphic Alternatives, Old and New." Proceedings of the SAS Global Forum 2012 Conference - Paper 235-2012. Cary, NC: SAS Institute Inc. Available at http://support.sas.com/resources/papers/proceedings12/235-2012.pdf.

Booth, Allison McMahill. 2010. "Evolve from a Carpenter's Apprentice to a Master Woodworker: Creating a Plan for Your Reports and Avoiding Common Pitfalls in REPORT Procedure Coding." Proceedings of the SAS Global Forum 2010 Conference - Paper 133-2010. Cary, NC: SAS Institute Inc. Available at http://support.sas.com/resources/papers/proceedings10/133-2010.pdf.

Booth, Allison McMahill. 2011. "Beyond the Basics: Advanced REPORT Procedure Tips and Tricks Updated for SAS® 9.2." Proceedings of the SAS Global Forum 2011 Conference - Paper 246-2011, Cary, NC: SAS Institute Inc. Available at http://support.sas.com/resources/papers/proceedings11/246-2011.pdf.

Booth, Allison McMahill. 2012. "PROC REPORT Unwrapped: Exploring the Secrets behind One of the Most Popular Procedures in Base SAS® Software." Proceedings of the 2012 PharmaSUG Conference - Paper TF20. Cary, NC: SAS Institute Inc. Available at http://www.pharmasug.org/proceedings/2012/TF/PharmaSUG-2012-TF20-SAS.pdf.

Campbell, Jillian. 2008. "Column Headings and Super-Headings: Using ACROSS Variables in PROC REPORT." Proceedings of the SAS Global Forum 2008 Conference - Paper 265-2008. Cary, NC: SAS Institute Inc. Available at http://www2.sas.com/proceedings/forum2008/265-2008.pdf.

Carpenter, Arthur L. 2006. "Advanced PROC REPORT: Traffic Lighting - Controlling Cell Attributes With Your Data." Available at http://www.caloxy.com/papers/69-TUT.pdf.

Carpenter, Arthur L. 2007. "Advanced PROC REPORT: Doing More in the Compute Block." Proceedings of the SAS Global Forum 2007 Conference - Paper 242 - 2007. Cary, NC: SAS Institute Inc. Available at http://www2.sas.com/proceedings/forum2007/242-2007.pdf.

Carpenter, Art. 2007. *Carpenter's Complete Guide to the SAS REPORT Procedure*. Cary, NC: SAS Institute Inc.

Carpenter, Arthur L. 2008. "PROC REPORT: Compute Block Basics – Part II Practicum." Proceedings of the SAS Global Forum 2008 Conference - Paper 188-2008. Cary, NC: SAS Institute Inc. Available at http://www2.sas.com/proceedings/forum2008/188-2008.pdf.

Carpenter, Arthur L. 2011. "PROC TABULATE: Doing More." Proceedings of the SAS Global Forum 2011 Conference - Paper 173-2011. Cary, NC: SAS Institute Inc. Available at http://support.sas.com/resources/papers/proceedings11/173-2011.pdf.

Chapman, David D. 2002. "Using PROC REPORT To Produce Tables With Cumulative Totals and Row Differences." Proceedings of the Twenty-Seventh Annual SAS Users Group International Conference - Paper 120-27. Cary, NC: SAS Institute Inc. Available at http://www2.sas.com/proceedings/sugi27/p120-27.pdf.

Croghan, Carry W. 2004. "PICTURE Perfect: In Depth Look at the PICTURE format." SESUG 2004: The Proceedings of the SouthEast SAS Users Group - Paper TU03. Cary, NC: SAS Institute Inc. Available at http://analytics.ncsu.edu/sesug/2004/TU03-Croghan.pdf.

DelGobbo, Vincent. 2007. "Creating Multi-Sheet Excel Workbooks the Easy Way with SAS®." Proceedings of the SAS Global Forum 2007 Conference - Paper 120-2007. Cary, NC: SAS Institute Inc. Available at http://support.sas.com/rnd/papers/sgf07/sgf2007-excel.pdf.

DelGobbo, Vincent. 2009. "More Tips and Tricks for Creating Multi-Sheet Microsoft Excel Workbooks the Easy Way with SAS®." Proceedings of the SAS Global Forum 2009 Conference - Paper 152-2009. Cary, NC: SAS Institute Inc. Available at http://support.sas.com/resources/papers/proceedings09/152-2009.pdf.

DelGobbo, Vincent. 2011. "Creating Stylish Multi-Sheet Microsoft Excel Workbooks the Easy Way with SAS®." Proceedings of the SAS Global Forum 2011 Conference - Paper 170-2011. Cary, NC: SAS Institute Inc. Available at http://support.sas.com/resources/papers/proceedings11/170-2011.pdf.

Ewing, Daphne, and Ray Pass. 2005. "So Now You're Using PROC REPORT–Is It Pretty and Automated?" Proceedings of the Thirtieth Annual SAS Users Group International Conference - Paper 244-30. Cary, NC: SAS Institute Inc. Available at http://www2.sas.com/proceedings/sugi30/244-30.pdf.

Fine, Lisa. 2011. "ORDER, ORDER PLEASE: SORTING DATA USING PROC REPORT." Proceedings of the SAS Global Forum 2011 Conference - Paper 090-2011. Cary, NC: SAS Institute Inc. Available at http://support.sas.com/resources/papers/proceedings11/090-2011.pdf.

Gebhart, Eric. 2010. "ODS ExcelXP: Tag Attr Is It! Using and Understanding the TAGATTR= Style Attribute with the ExcelXP Tagset." Proceedings of the SAS Global Forum 2010 Conference - Paper 031-2010. Cary, NC: SAS Institute Inc. Available at http://support.sas.com/resources/papers/proceedings10/031-2010.pdf.

Haworth, Lauren. 2005. "SAS® with Style: Creating Your Own ODS Style Template for PDF Output." Proceedings of the Thirtieth Annual SAS Users Group International Conference - Paper 132-30. Cary, NC: SAS Institute Inc. Available at http://www2.sas.com/proceedings/sugi30/132-30.pdf.

Haworth, Lauren. 2006. "PROC TEMPLATE: The Basics." Proceedings of the Thirty-First Annual SAS Users Group International Conference - Paper 112-31. Cary, NC: SAS Institute Inc. Available at http://www2.sas.com/proceedings/sugi31/112-31.pdf.

Haworth, Lauren. 2009. "Advanced Topics in ODS." SESUG 2009: The Proceedings of the SouthEast SAS Users Group - Paper RV-010. Cary, NC: SAS Institute Inc. Available at http://analytics.ncsu.edu/sesug/2009/RV010.Haworth.pdf.

Haworth, Lauren E., Cynthia L. Zender, and Michele M. Burlew. 2009. *Output Delivery System: The Basics and Beyond*. Cary, NC: SAS Institute Inc.

Haworth, Lauren. 2011. "ODS RTF: the Basics and Beyond." Proceedings of the SAS Global Forum 2011 Conference - Paper 263-2011. Cary, NC: SAS Institute Inc. Available at http://support.sas.com/resources/papers/proceedings11/263-2011.pdf.

Hunt, Stephen. 2006. "Say 'Hello' to #BYVAL: Re-introducing a Hot Little #." Proceedings of the 2006 PharmaSUG Conference - Paper TT01. Cary, NC: SAS Institute Inc. Available at http://www.lexjansen.com/pharmasug/2006/TechnicalTechniques/TT01.pdf.

Huntley, Scott. 2006. "Let the ODS PRINTER Statement Take Your Output into the Twenty-First Century." Proceedings of the Thirty-First Annual SAS Users Group International Conference - Paper 227-31. Cary, NC: SAS Institute Inc. Available at http://www2.sas.com/proceedings/sugi31/227-31.pdf.

Karp, Andrew H. 2006. "Getting in to the Picture (Format)." Proceedings of the Thirty-First Annual SAS Users Group International Conference - Paper 243-31. Cary, NC: SAS Institute Inc. Available at http://www2.sas.com/proceedings/sugi31/243-31.pdf.

Karp, Andrew H. 2009. "Traffic-Lighting Your Reports the Easy Way with PROC REPORT and ODS." Proceedings of the SAS Global Forum 2009 Conference - Paper 273-2009. Cary, NC: SAS Institute Inc. Available at http://support.sas.com/resources/papers/proceedings09/273-2009.pdf.

Kincaid, Chuck. 2010. "SGPANEL: Telling the Story Better". Proceedings of the SAS Global Forum 2010 Conference - Paper 234-2010. Cary, NC: SAS Institute Inc. Available at http://support.sas.com/resources/papers/proceedings10/234-2010.pdf.

Lafler, Kirk Paul. 2009. "Exploring PROC SQL® Joins and Join Algorithms." Proceedings of the SAS Global Forum 2009 Conference - Paper 035-2009. Cary, NC: SAS Institute Inc. Available at http://support.sas.com/resources/papers/proceedings09/035-2009.pdf.

Lavery, Russell. 2003. "An Animated Guide©: Proc Report: The File Behind the Scenes." SESUG 2003: The Proceedings of the 2003 SouthEast SAS Users Group Conference - Paper TU-08. Cary, NC: SAS Institute Inc. Available at http://analytics.ncsu.edu/sesug/2003/TU08-Lavery.pdf.

Lawhorn, Bari. 2011. "Let's Give 'Em Something to TOC About: Transforming the Table of Contents of Your PDF File." SESUG 2011: The Proceedings of the SouthEast SAS Users Group -

Paper SS-03. Cary, NC: SAS Institute Inc. Carolina State University. Available at http://analytics.ncsu.edu/sesug/2011/SS03.Lawhorn.pdf.

Lund, Pete. 2001. "More than Just Value: A Look Into the Depths of PROC FORMAT." Proceedings of the Twenty-Sixth Annual SAS Users Group International Conference - Paper 18-26. Cary, NC: SAS Institute Inc. Available at http://www2.sas.com/proceedings/sugi26/p018-26.pdf.

Lund, Pete. 2011. "You Did That Report in SAS®!?: The Power of the ODS PDF Destination." Proceedings of the SAS Global Forum 2011 Conference - Paper 247-2011. Cary, NC: SAS Institute Inc. Available at http://support.sas.com/resources/papers/proceedings11/247-2011.pdf.

McMahill, Allison. 2007. "Beyond the Basics: Advanced PROC REPORT Tips and Tricks." Proceedings of the SAS Global Forum 2007 Conference - Paper 276-2007. Cary, NC: SAS Institute Inc. Available at http://www2.sas.com/proceedings/forum2007/276-2007.pdf.

McRitchie, David F. 2006. "Font Tables as Rendered by Your Browser." Available at http://dmcritchie.mvps.org/rexx/htm/fonts.htm.

O'Connor, Daniel, and Scott Huntley. 2009. "Breaking New Ground with SAS® 9.2 ODS Layout Enhancements." Proceedings of the SAS Global Forum 2009 Conference - Paper 043-2009. Cary, NC: SAS Institute Inc. Available at http://support.sas.com/resources/papers/proceedings09/043-2009.pdf.

Parker, Chevell. 2011. "The Perfect Marriage: The SAS® Output Delivery System (ODS) and Microsoft Office." Proceedings of the SAS Global Forum 2011 Conference - Paper 250-2011. Cary, NC: SAS Institute Inc. Available at http://support.sas.com/resources/papers/proceedings11/250-2011.pdf.

Pass, Ray, and Sandy McNeill. 2004. "PROC TABULATE: Doin' It in Style!" Proceedings of the Twenty-Ninth Annual SAS Users Group International Conference - Paper 085-29. Cary, NC: SAS Institute Inc. Available at http://www2.sas.com/proceedings/sugi29/085-29.pdf.

SAS Institute Inc. 2007. SAS® 9 ODS RTF Tip Sheet. Cary, NC: SAS Institute Inc. Available at http://support.sas.com/rnd/base/ods/odsrtf/rtf-tips.pdf.

SAS Institute Inc. 2008. SAS® 9 ODS TAGSETS.RTF Tip Sheet. Cary, NC: SAS Institute Inc. Available at http://support.sas.com/rnd/base/ods/odsrtf/rtf-tagset-tips.pdf.

SAS Institute Inc. 2009. KNOWLEDGE BASE / SAMPLES & SAS NOTES. "Sample 36288: Repeating text on an RTF page when the BODYTITLE option is in effect." Cary, NC: SAS Institute Inc. Available at http://support.sas.com/kb/36/288.html.

SAS Institute Inc. 2009. KNOWLEDGE BASE / SAMPLES & SAS NOTES. "Usage Note 15883: Length limitations when submitting SAS code." Cary, NC: SAS Institute Inc. Available at http://support.sas.com/kb/15/883.html.

SAS Institute Inc. 2011. KNOWLEDGE BASE / SAMPLES & SAS NOTES. "Sample 43765: How to use an array in a COMPUTE block with PROC REPORT." Cary, NC: SAS Institute Inc. Available at http://support.sas.com/kb/43/765.html.

SAS Institute Inc. 2012. SAS® 9.2 KNOWLEDGE BASE / PRODUCT DOCUMENTATION / SAS 9.2 Documentation. "How PROC REPORT Builds a Report." Cary, NC: SAS Institute Inc. Available at http://support.sas.com/documentation/cdl/en/proc/61895/HTML/default/viewer.htm#a002473631.htm.

SAS Institute Inc. 2013. *Base SAS® 9.3 Procedures Guide, Second Edition*. Cary, NC: SAS Institute Inc. Available at http://support.sas.com/documentation/cdl/en/proc/65145/HTML/default/viewer.htm#titlepage.htm.

SAS Institute Inc. 2013. KNOWLEDGE BASE/ FOCUS AREAS. "Base SAS Try This Demo: The ExcelXP Tagset and Microsoft Excel." Cary, NC: SAS Institute Inc. Available at http://support.sas.com/rnd/base/ods/odsmarkup/excelxp_demo.html.

SAS Institute Inc. 2013. KNOWLEDGE BASE/ FOCUS AREAS. "Base SAS Quick Reference for the TAGSETS.EXCELXP Tagset." Cary, NC: SAS Institute Inc. Available at http://support.sas.com/rnd/base/ods/odsmarkup/excelxp_help.html.

SAS Institute Inc. 2013. KNOWLEDGE BASE FOCUS AREAS. "Base SAS Enhancements to ODS RTF for SAS 9.2." Cary, NC: SAS Institute Inc. Available at http://support.sas.com/rnd/base/new92/92rtf.html#inline.

SAS Institute Inc. 2013. KNOWLEDGE BASE / PAPERS, TS-688, "Defining Colors Using Hex Values." Cary, NC: SAS Institute Inc. Available at http://support.sas.com/techsup/technote/ts688/ts688.html.

SAS Institute Inc. 2013. SAS® 9.2 KNOWLEDGE BASE / PRODUCT DOCUMENTATION / BASE SAS 9.2. Cary, NC: SAS Institute Inc. Available at http://support.sas.com/documentation/onlinedoc/base/index.html#base92.

SAS Institute Inc. 2013. KNOWLEDGE BASE / SAMPLES & SAS NOTES. "Sample 49590: Insert special symbols as a table value in ODS MARKUP destinations." Cary, NC: SAS Institute Inc. Available at http://support.sas.com/kb/49/590.html.

SAS Institute Inc. 2013. *SAS® 9.3 Output Delivery System: User's Guide, Second Edition*. "TEMPLATE Procedure: Creating a Style Template." Cary, NC: SAS Institute Inc. Available at http://support.sas.com/documentation/cdl/en/odsug/65308/HTML/default/viewer.htm#n19a4b40swc766n18qczor47r08f.htm.

Sembongi, Yumi. 2010. "PROC REPORT COMPUTE Block and Conditional Footnote." Proceedings of the 2010 PharmaSUG Conference - Paper PO06. Cary, NC: SAS Institute Inc. Available at http://www.lexjansen.com/pharmasug/2010/po/po06.pdf.

Slaughter, Susan J. and Lora D. Delwiche. 2010. "Using PROC SGPLOT for Quick High-Quality Graphs." Proceedings of the SAS Global Forum 2010 Conference - Paper 154-2010. Cary, NC: SAS Institute Inc. Available at http://support.sas.com/resources/papers/proceedings10/154-2010.pdf.

Smith, Kevin D. 2006. The TEMPLATE Procedure Styles: Evolution and Revolution Proceedings of the Thirty-First Annual SAS Users Group International Conference - Paper 053-31. Cary, NC: SAS Institute Inc. Available at http://www2.sas.com/proceedings/sugi31/053-31.pdf.

Smith, Kevin D. 2013. *PROC TEMPLATE Made Easy: A Guide for SAS® Users*. Cary, NC: SAS Institute Inc.

Zender, Cynthia L. 2007. "Funny ^Stuff~ in My Code: Using ODS ESCAPECHAR." Proceedings of the SAS Global Forum 2007 Conference - Paper 099-2007. Cary, NC: SAS Institute Inc. Available at http://www2.sas.com/proceedings/forum2007/099-2007.pdf.

Zender, Cynthia L. 2008. "Creating Complex Reports." Proceedings of the SAS Global Forum 2008 Conference - Paper 173-2008. Cary, NC: SAS Institute Inc. Available at http://www2.sas.com/proceedings/forum2008/173-2008.pdf.

Zender, Cynthia L. 2010. "Tiptoe Through the Templates." Proceedings of the SAS Global Forum 2011 Conference - Paper 227-2009. Cary, NC: SAS Institute Inc. Available at http://support.sas.com/resources/papers/proceedings09/227-2009.pdf.

Zender, Cynthia L. 2011. "Don't Gamble with Your Output: How to Use Microsoft Formats with ODS." Proceedings of the SAS Global Forum 2011 Conference - Paper 266-2011. Cary, NC: SAS Institute Inc. Available at http://support.sas.com/resources/papers/proceedings11/266-2011.pdf.

DATA SETS

SAS Institute Inc. 2013. SAS®, SASHELP.HEART, Cary, NC: SAS Institute Inc.

SAS Institute Inc. 2013. SAS®, SASHELP.SNACKS, Cary, NC: SAS Institute Inc.

SAS Institute Inc. 2013. SAS®, SASHELP.IRIS, Cary, NC: SAS Institute Inc.

SAS Institute Inc. 2013. SAS®, SASHELP.SHOES, Cary, NC: SAS Institute Inc.

SAS Institute Inc. 2013. SAS®, SASHELP.CARS, Cary, NC: SAS Institute Inc.

"Copyright 2013, SAS Institute Inc., Cary, NC, USA. All Rights Reserved. Reproduced with permission of SAS Institute Inc., Cary, NC"

Index

A

ABAR statement 148
ABSOLUTE_COLUMN_WIDTH 151–152
ACROSS option 229
  See also Weekly Sales report
ACROSS variable 135–136, 173, 174
ANALYSIS variable 36, 196, 211
ASIS=ON style attribute 195
assigning report order to variables 27–28
asterisk ("*") 172, 177
attributes
  ASIS=ON 195
  POSTIMAGE= 205, 208
  PREIMAGE= 196, 206, 208, 211
  STYLE= 208
  TAGATTR 148, 154
autofilters 143, 153

B

BANKER template 208
bar charts, creating with SGPANEL procedure 230–234
BARWIDTH option 229
Base SAS® 9.3 Procedures Guide, 2nd Edition 101
"Before Formatting" program 49–52
BLANK variable 42–43, 62–64
BODYTITLE option 17, 195
BREAK AFTER statement 173
BREAK statement 22, 38t, 53, 55, 65, 66f, 114
BY statement 173, 251
BY VARIABLES 240
ByStatusALL worksheet 149–153
ByStatusCOL worksheet 144–148
ByStatusROW worksheet 144–148
#BYVAL option 254
#BYVAR option 254

C

calendar grid, obtaining and merging with sales 162–166
CALL DEFINE statement 37, 58, 59, 66f, 99, 102
caret ("^") 13
categorical counts 76–81, 137–141
categorical variables, in Demographic and Baseline Characteristic Report 75–81
CATNAME variable 36
CATORD variable 27, 28t, 29, 36, 38t
cell borders, adding to National Sales Report 59–62
CH1Setup data set 8–14
CH2Sales data set 46
Ch3Demo data set 73
Ch4Lesn data set 99–100
CH5Tgxml data set 134
CH6CAL data set 160
Ch8Graph data set 219
Ch9Stat data set 242
CLASS FONTS statement 10–11
CLASS HEADER statement 11
CLASS statements 47–48, 79, 130
CLASS SYSTEMFOOTER statement 12
CLASS SYSTEMTITLE statement 12
CLASS TABLE statement 11
closing ExcelXP workbook 142–143
code
  for closing ExcelXP workbook 142–143
  for combining results 85–88
  for creating formats and informats 134–137
  for creating horizontal bar charts 230–231
  for creating ordered variables 28–29
  for creating SGPLOT vertical bar charts 227–230
  for creating Summary Data Set 220–222

264 Index

for displaying images above body of table 210–211
for displaying images as column headers 204–206
for displaying images in page titles 207–208
for displaying potential data issues 108–119
for displaying regions in National Sales Report 54
for footnotes 30
for inserting arrows in National Sales Report 56–59
for obtaining categorical counts and percentages 76–81, 137–141
for obtaining images as columns of data 193–197
for obtaining population counts 75
for obtaining regional ranking information 222–223
for obtaining statistics 242–245
for ODS RTF 30
for opening ExcelXP workbook 142–143
pre-processing for Summary report 24–27
for printing MSRP Comparison report 245–254
for producing ByStatusALL worksheet 149–153
for producing ByStatusCOL worksheet 144–148
for producing ByStatusROW worksheet 144–148
for producing Weekly Sales report 167–177
for repeating images above and below tables 198–202
for setting up initial options for ExcelXP workbook 142–143
for store and branch display in National Sales Report 55
for Summary Report 34–35
for titles 30
ColDtCt 114, 118, 118–119*t*
ColFewer 117*t*
ColMore 117*t*

ColSubj 105
ColSzDf 114, 118, 118–119*t*
COLUMN statement
    adding new report variables to 109–110
    in complementary reports 22, 36
    creating spanning headers 119–120
    in embedding images 196, 205
    in highly detailed reports 56, 57, 58, 59–62, 63, 66f
    in Lesion Data Quality report 104
    in multi-sheet workbooks 152
    in Summary statistics 251, 252
COLUMN statement variables
    See REPORT variables
columns
    adding to National Sales Report 62–64
    displaying images as headers for 204–206
    obtaining images as 193–197
COLUMNS= option 226
complementary reports
    about 2
    Detail report 15–17
    examples 2–39
    goals for creating 4–6
    implementing 8–14
    ODS style template used for 6, 7–8
    producing Detail report with REPORT procedure 17–23
    producing Summary Report with REPORT procedure 6, 31–39
    programs used for 8
    source data for 6–7
    writing Detail Report program 14–15
    writing Summary Report program 23–30
COMPUTE BEFORE statement 37, 38*t*, 173
COMPUTE block 53, 56, 59, 63, 66, 66*f*, 67*f*, 99, 114, 240, 254
COMPUTE Block LINE statements 35, 37, 38*t*, 90
COMPUTE block variables 101–102, 120, 196
COMPUTE variable 252
COMPUTE_BEFORE_PAGE 253
COMPUTED variable 66*f*, 99, 102, 114

continuous variables, in Demographic and Baseline Characteristic report 81–85
counts, obtaining 137–141
"Creating Stylish Multi-Sheet Microsoft Excel Workbooks the Easy Way with SAS® (2011)" (DelGobbo) 151
CUMFIN 221

D

Dashboard Report of Shoe Sales
    about 216
    creating formats needed for outputs 226–227
    creating new ODS style templates 223–226
    creating ODS layout 226
    example 216–235
    goals for creating 218
    implementing 220–222
    obtaining regional ranking information 222–223
    ODS style template used for 219
    programs used for 219
    source data for 218–219
    using REPORT procedure to obtain tabular output 231–232
    using SGPANEL procedure to create bar charts 232–234
    using SGPLOT procedure to create horizontal bar charts 230–231
    using SGPLOT procedure to create vertical bar charts 227–230
data
    identifying potential issues with 107–119
    pre-processing 4
DATA step variables 62, 67f, 87, 101–102, 105, 110, 113, 116, 120
DATALABEL option 229, 231, 234
DATALABELATTRS= option 234
DBAR statement 148
&DEBUG macro 83
DEFINE statement
    adding for new report variables 110, 113

    in complementary reports 22
    descriptions of 174t
    in embedding images 196
    in highly detailed reports 53, 54–55, 56, 57, 58, 59–62, 64, 66f, 67f
    in Lesion Data Quality report 104–105
    in multi-sheet workbooks 153
    in reporting metrics 91
    in Summary statistics 243–244, 252, 253
    in Weekly Sales report 173
DelGobbo, Vince
    "Creating Stylish Multi-Sheet Microsoft Excel Workbooks the Easy Way with SAS® (2011)" 151
Demographic and Baseline Characteristic Report
    about 70–71
    categorical variables in 75–81
    continuous variables in 81–85
    creating final tables 85–88
    goals for 72
    implementing 74
    obtaining population counts 74–75
    ODS style template used for 74
    producing with REPORT procedure 89–91
    programs used for 74
    source data for 73
DESCRIP column 202, 206
Detail report 15–17
digit selectors 171
displaying
    images above body of table 208–211
    images as column headers 203–206
    images in page titles 206–208
    regions in National Sales Report 53–54
    store and branch column data in National Sales Report 54–55
    watermarks on reports 212–213
DO loop 164
DSTOTAL variable 176
DtaBlCt 113
DtaDtCt 113, 114, 118
DtaFDt 113, 114, 118
DtaLesCt 113, 114

266 *Index*

DtaSize 113, 114, 118
DtaSubCt 105, 113
DtaSubj 105

E

ELSE IF statement 253
embedded titles 143
embedding images
    about 182
    displaying images above body of table 208–211
    displaying images as column headers 203–206
    displaying images in page titles 206–208
    displaying watermarks on reports 212–213
    example 182–214
    goals for 188
    implementing 190–192
    obtaining images as columns of data 192–197
    ODS style template used for 190
    program setup code for 191–192
    programs used for 190
    repeated images above and below tables 197–202
    source data for 188–189
ExcelXP workbook 142–143

F

file paths 190
FILLATTRS= option 231
FINAMT 221
$FLOWER format 192, 196
FOOTERY option 17
footnotes, code for 30
FORMAT procedure 87
formats 134–137, 226–227
FREQ procedure 74, 75, 137–141
frozen headers 143
functions

INTNX 164
PUT 84
REPEAT 148
ROUND 84, 172

G

goals
    for creating complementary reports 4–6
    for creating Dashboard Report for Shoe Sales 218
    for creating Lesion Data Quality report 98–99
    for creating multi-sheet workbooks 128–129
    for creating Weekly Sales report 158
    for Demographic and Baseline Characteristic Report 72
    for embedding images 188
    for formatting National Sales Report 45–46
    for MSRP Comparison report 240
GraphData1 225*t*
GraphData2 225*t*
GraphDataFont 225*t*
GraphLabelFont 226*t*
GraphTitleFont 226*t*
GraphValueFont 225*t*
GROUP option 36–37, 102
GROUP variable 173

H

HARVEST template 242
HBAR statement 231
HEIGHT= option 195, 228
Hex Code 58
highly detailed reports
    See National Sales Report
horizontal bar charts 230–231

I

IF statement 253
image file names 190
IMAGEFMT= option 228
IMAGENAME= option 228
images
    See embedding images
implementing
    complementary reports 8–14
    Dashboard Report of Shoe Sales 220–222
    Demographic and Baseline Characteristic Report 74
    embedding images 190–192
    formatting National Sales Report 53
    Lesion Data Quality report 101–102
    MSRP Comparison report 242–245
    multi-sheet workbook 134–137
    Weekly Sales report 160–162
&INDATA macro 83
informats 134–137
INTNX function 164
Iris City Gardens 188
IRIS data set 199–202

K

KEYLEGEND statement 229

L

LAYOUT START statement 226
LAYOUT=COLUMNLATTICE option 233
Lesion Data Quality report
    about 96
    creating spanning headers for 119–120
    example 96–121
    goals for creating 98–99
    identifying potential data issues 107–119
    implementing 101–102
    ODS style template used for 100–101
    ORDER by 102–107
    Print Subject ID 102–107
    programs used for 101
    source data for 99–100
LINE statement 148
lines, adding to National Sales Report 65–66

M

%MACRO TAB 77
macros
    &DEBUG 83
    &INDATA 83
    %MACRO TAB 77
    MEANS 81–85
    &MEANVAR 83
    OUTPATH 192
    PREPROC 13–14, 14t
    &RAWDEC 83, 84
    &RNDDEC 83
mapping variables/values to one column 23–24
McCullough, Greg 188
MEADOW template 219, 223–226, 235
MEADOWG template 219
MEANS macro 81–85
MEANS procedure 72, 74, 83, 84, 85–88
&MEANVAR macro 83
message characters 171
metrics, reporting 70
    See also Demographic and Baseline Characteristic Report
MISSING option 79
MSRP Comparison report
    about 238
    examples 238–254
    goals for 240
    implementing 242–245
    ODS style template used for 242
    printing 245–254
    producing 245–254
    programs used for 242
    source data for 240–242
MSRPORD alias 251

268 *Index*

multi-sheet workbook
   about 124
   example 124–154
   goals for creating 128–129
   implementing 134–137
   obtaining counts and percentages 137–141
   ODS style template used for 130–133
   producing 141–154
   producing with REPORT procedure and ODS TAGSET 141–143
   programs used for 134
   source data for 129–130

N

$NAME format 153
National Sales Report
   about 42
   adding blank columns to 62–64
   adding blank lines to 65–66
   adding bottom cell borders to 59–62
   adding spanning headers to 59–62
   adding underlines to 59–62
   "Before Formatting" program 49–52
   displaying regions in 53–54
   displaying store and branch column data 54–55
   example 42–67
   goals for 45–46
   implementing formatting of 53
   inserting arrows in 56–59
   ODS style template used for 46–47
   programs used for 47
   source data for 46
   TEMPLATE procedure program to create new style template 47–49
NOBYLINE option 173, 249
NODATE option 190, 192
NONUMBER option 190
NOPRINT option 38*t*, 53, 54, 66*f*, 85–87, 104–105, 116, 148, 173, 211, 232
NOPRINTED variable 196
NOTOC option 249

NOVARNAME option 234
NUMDAY variable 176

O

ODS Close statement 143
ODS escape character ("^") 190, 191, 213*t*
ODS GRAPHICS option 228, 230, 233
ODS Journal style template 6, 7–8
ODS layout, creating 226
ODS LAYOUT END statement 226, 234
ODS LAYOUT START statement 226
ODS PROCLABEL statement 251, 254
ODS RTF statement 30, 190, 195
ODS style template
   for creating Lesion Data Quality report 100–101
   creating new 223–226
   for Dashboard Report of Shoe Sales 219
   for embedding images 190
   used for complementary reports 6, 7–8
   used for creating multi-sheet workbooks 130–133
   used for Demographic and Baseline Characteristic Report 74
   used for formatting National Sales Report 46–47
   used for MSRP Comparison report 242
   for Weekly Sales report 160
ODS TAGSET statement 141–143, 148
opening ExcelXP workbook 142–143
options
   ACROSS 229
   to apply to all worksheets 143*t*
   BARWIDTH 229
   BODYTITLE 17, 195
   #BYVAL 254
   #BYVAR 254
   COLUMNS= 226
   DATALABEL 229, 231, 234
   DATALABELATTRS= 234
   FILLATTRS= 231
   FOOTERY 17

GROUP 36–37, 102
HEIGHT= 195, 228
IMAGEFMT= 228
IMAGENAME= 228
LAYOUT=COLUMNLATTICE 233
MISSING 79
NOBYLINE 173, 249
NODATE 190, 192
NONUMBER 190
NOPRINT 38t, 53, 54, 66f, 85–87, 104–105, 116, 148, 173, 211, 232
NOTOC 249
NOVARNAME 234
ODS GRAPHICS 228, 230, 233
ORDER 22, 53, 54, 55, 99, 102, 240, 252
PDFTOC 254
PRETEXT= 205
RESET 228
ROWS= 226
SPACING=48 233
SPANROWS 240, 251, 254
STARTPAGE 202
SUBJECT 104–105
SUMMARIZE 22, 38t, 51
SUPPRESS 51
UNIFORM 249
WATERMARK= 213
WIDTH= 228
ORDER option 22, 53, 54, 55, 99, 102, 240, 252
ORIGIN variable 248–249
OUTPATH macro variable 192
output data set (PROUT) 202, 203f

P

PAGETIT parameter 147
PANELBY statement 233
PARENT= statement 130
PCT parameter 147
PCTC character string 79
PCTDEC picture format 227, 231
PDFTOC option 254
percentages, obtaining 76–81, 137–141

PICTURE format 75, 76
picture formats 166, 171, 172t
pipe ("|") character 211
population counts, obtaining for Demographic and Baseline Characteristic Report 74–75
POSTIMAGE= attribute 205, 208
PREIMAGE= attribute 196, 206, 208, 211
PRELOADFMT 76–77, 79
PREPROC macro 13–14, 14t
pre-processing
    code for Summary report 24–27
    data 4
PRETEXT= option 205
PRINT procedure 13–14, 14t, 25, 79–81, 84, 85t, 115t
procedures
    See also REPORT procedure
    FORMAT 87
    FREQ 74, 75, 137–141
    MEANS 72, 74, 83, 84, 85–88
    PRINT 13–14, 14t, 25, 79–81, 84, 85t, 115t
    RANK 218, 222–223, 234
    SGPANEL 232–234, 235
    SGPLOT 227–231, 235
    TABULATE 72, 74, 75–76, 79, 84, 85–88
    TEMPLATE 12–13, 47–52
    TRANSPOSE 24–27
PRODUCT variable 176
programs
    used for complementary reports 8
    used for creating Lesion Data Quality report 101
    used for Dashboard Report of Shoe Sales 219
    used for Demographic and Baseline Characteristic Report 74
    used for embedding images 190
    used for MSRP Comparison report 242
    used for multi-sheet workbooks 134
    used for National Sales Report 47
    used for Weekly Sales Report 160
PROUT (output data set) 202, 203f

PRRPT1 macro 147
PRSSUM data set 37, 38t
PUT function 84
%PUT statement 75

Q

QUARTILES data set 242–245

R

RANK procedure 218, 222–223, 234
&RAWDEC macro 83, 84
RBREAK BEFORE statement 221
REGION DEFINE statement 66f
Region display (National Sales Report) 53–54
regional ranking information, obtaining for Demographic and Baseline Characteristic Report 222–223
REPEAT function 148
REPNUM parameter 147
REPORT procedure
    See also specific topics
    about 14–19
    code for 19–20
    obtaining statistics with 242–245
    obtaining tabular output with 231–232
    producing Demographic and Baseline Characteristic report with 89–91
    producing Detail report with 17–23
    producing multi-sheet workbooks with 141–143
    producing Summary Report with 31–39
    using to clarify COMPUTE block operations 99
REPORT variables 101–102, 120
RESET option 228
RESPONSE variable 229, 231, 234
RETAMT picture format 227
&RNDDEC macro 83
ROUND function 84, 172

ROWS= option 226
"RSTYLERTF" template 46–47

S

SALES data set 220–222, 232
SASHELP.CARS data set 240–242
SASHELP.HEART data set 129–130
SASHELP.IRIS data set 188–189
SASHELP.SHOES data set 218–219
SASWEB ODS style template 160–162, 195, 205
SASWEB template 173
SASWEBR template 160–162
setting initial options for ExcelXP workbook 142–143
setup options 8, 190
SGPANEL procedure 232–234, 235
SGPLOT procedure 227–231, 235
SHEET parameter 147
source data
    for complementary reports 6–7
    for creating Lesion Data Quality report 99–100
    for Dashboard Report of Shoe Sales 218–219
    for Demographic and Baseline Characteristic Report 73
    for embedding images 188–189
    for formatting National Sales Report 46
    for MSRP Comparison report 240–242
    for multi-sheet workbooks 129–130
    for Weekly Sales report 158–160
SPACING=48 option 233
spanning headers
    adding to National Sales Report 59–62
    creating 119–120
    inserting images as 204–205
SPANROWS option 240, 251, 254
SPECIES column 202, 211
SPECIES DEFINE statement 192, 213t
STARTPAGE option 202

statements
    See also COLUMN statement; DEFINE statement
    BY 173, 251
    ABAR 148
    BREAK 22, 38t, 53, 55, 65, 66f, 114
    BREAK AFTER 173
    CALL DEFINE 37, 58, 59, 66f, 99, 102
    CLASS 47–48, 79, 130
    CLASS FONTS 10–11
    CLASS HEADER 11
    CLASS SYSTEMFOOTER 12
    CLASS SYSTEMTITLE 12
    CLASS TABLE 11
    COMPUTE BEFORE 37, 38t, 173
    COMPUTE Block LINE 35, 37, 38t, 90
    DBAR 148
    ELSE IF 253
    HBAR 231
    IF 253
    KEYLEGEND 229
    LAYOUT START 226
    LINE 148
    ODS Close 143
    ODS LAYOUT END 226, 234
    ODS LAYOUT START 226
    ODS PROCLABEL 251, 254
    ODS RTF 30, 190, 195
    ODS TAGSET 141–143, 148
    PANELBY 233
    PARENT= 130
    %PUT 75
    RBREAK BEFORE 221
    REGION DEFINE 66f
    SPECIES DEFINE 192, 213t
    SUBJECT DEFINE 102
    TAGSETS.RTF 190
    TITLE 190, 208
    VBAR 228, 234
STATISTICAL template 130–133
STATISTICALX template 130–133
statistics, obtaining 242–245
style, applying overrides to 22

STYLE= attribute 208
STYLE template 20
SUBCAT variable 81
SUBCTORD variable 27, 28t, 29, 36, 38t, 88
SUBJECT DEFINE statement 102
SUBJECT option 104–105
SUMMARIZE option 22, 38t, 51
Summary data set 220–222
Summary report
    code for 34–35
    pre-processing data for 24–27
    producing with REPORT procedure 6, 31–39
    titles, footnotes, and ODS RTF preparation 30
    writing program for 23–30
summary statistics
    See MSRP Comparison report
SUPPRESS option 51

T

table of contents (TOC) 249
tabular data
    See Dashboard Report of Shoe Sales
TABULATE procedure 72, 74, 75–76, 79, 84, 85–88
TAGATTR style attribute 148, 154
TAGSETS.RTF statement 190
TEMPLATE procedure 12–13, 47–52
templates
    BANKER 208
    HARVEST 242
    MEADOW 219, 223–226, 235
    MEADOWG 219
    ODS Journal style 6, 7–8
    ODS style template (See ODS style template)
    "RSTYLERTF" 46–47
    SASWEB 173
    SASWEB ODS style 160–162, 195, 205
    SASWEBR 160–162
    STATISTICAL 130–133

STATISTICALX 130–133
STYLE 20
temporary variables
    See DATA step variables
TEXT0 variable 253
TEXT1...TEXT4 variables 253
TEXT5 variable 253–254
TEXT6 variable 253–254
TITLE statement 190, 208
Title_Footnote_Width 152
titles
    code for 30
    displaying images in 207–208
    embedded 143
TOC (table of contents) 249
TRANSPOSE procedure 24–27
TYPE2 alias 251

U

underlines, adding to National Sales Report 59–62
UNIFORM option 249

V

values, mapping to one column 23–24
$VAR 91
variables
    ACROSS 135–136, 173, 174
    adding to COLUMN statement 109–110
    ANALYSIS 36, 196, 211
    assigning report order to 27–28
    BLANK 42–43, 62–64
    categorical 75–81
    CATNAME 36
    CATORD 27, 28t, 29, 36, 38t
    code for creating ordered 28–29
    COMPUTE 252
    COMPUTE block 101–102, 120, 196
    COMPUTED 66f, 99, 102, 114
    continuous 81–85
    DATA step 62, 67f, 87, 101–102, 105, 110, 113, 116, 120
    DSTOTAL 176
    GROUP 173
    mapping to one column 23–24
    NOPRINTED 196
    NUMDAY 176
    ordered 28–29
    ORIGIN 248–249
    OUTPATH 192
    PRODUCT 176
    REPORT 101–102, 120
    RESPONSE 229, 231, 234
    SUBCAT 81
    SUBCTORD 27, 28t, 29, 36, 38t, 88
    TEXT0 253
    TEXT1...TEXT4 253
    TEXT5 253–254
    TEXT6 253–254
    VARNAME 76, 81, 84, 88, 88t, 91
    VARORD 88
VARNAME variable 76, 81, 84, 88, 88t, 91
VARORD variable 88
VBAR statement 228, 234
vertical bar charts 227–230

W

WATERMARK= option 213
watermarks, displaying on reports 212–213
Weekly Sales report
    about 156
    example 156–179
    goals for creating 158
    implementing 160–162
    obtaining calendar grid and merging with sales 162–166
    ODS style template used for 160
    place holders 177–178
    producing 166–177
    programs used for 160
    source data for 158–160
WIDTH= option 228

WIDTH_FUDGE 152
WIDTH_POINTS 151
Wingdings 58

X

XColDtCt 118
XColFDt 118
XColSize 118

Made in the USA
Middletown, DE
29 October 2014